Famous Scientific Illusions

Famous Scientific Illusions

By Nikola Tesla

Wilder Publications, LLC.
PO Box 3005
Radford VA 24143-3005
www.wilderpublications.com

ISBN 10: 1-934451-99-1
ISBN 13: 978-1-934451-99-1

First Edition

10 9 8 7 6 5 4 3 2 1

Table of Contents

Famous Scientific Illusions
Electrical Experimenter — February, 1919

The human brain, with all its wonderful capabilities and power, is far from being a faultless apparatus. Most of its parts may be in perfect working order, but some are atrophied, undeveloped or missing altogether. Great men of all classes and professions-scientists, inventors, and hard-headed financiers — have placed themselves on record with impossible theories, inoperative devices, and unrealizable schemes. It is doubtful that there could be found a single work of any one individual free of error. There is no such thing as an infallible brain. Invariably, some cells or fibers are wasting or unresponsive, with the result of impairing judgment, sense of proportion, or some other faculty. A man of genius eminently practical, whose name is a household word, has wasted the best years of his life in a visionary undertaking. A celebrated physicist was incapable of tracing the direction of an electric current according to a childishly simple rule. The writer, who was known to recite entire volumes by heart, has never been able to retain in memory and recapitulate in their proper order the words designating the colors of the rainbow, and can only ascertain them after long and laborious thought, strange as it may seem.

Our organs of reception, too, are deficient and deceptive. As a semblance of life is produced by a rapid succession of inanimate pictures, so many of our perceptions are but trickery of the senses, devoid of reality. The greatest triumphs of man were those in which his mind had to free itself from the influence of delusive appearances. Such was the revelation of Buddha that self is an illusion caused by the persistence and continuity of mental images: the discovery of Copernicus that contrary to all observation, this planet rotates around the sun; the recognition of Descartes that the human being is an automaton, governed by external influence and the idea that the earth is spherical, which led Columbus to the finding of this continent. And tho the minds of individuals supplement one another and science and experience are continually eliminating fallacies and misconceptions, much of our present knowledge is still incomplete and unreliable. We have sophisms in mathematics which cannot be disproved. Even in pure reasoning, free of the shortcomings of symbolic processes, we are often arrested by doubt which the strongest intelligences have been unable to dispel. Experimental science itself, most positive of all, is not unfailing.

In the following I shall consider three exceptionally interesting errors in the interpretation and application of physical phenomena which have for years dominated the minds of experts and men of science.

I. The Illusion of the Axial Rotation of the Moon

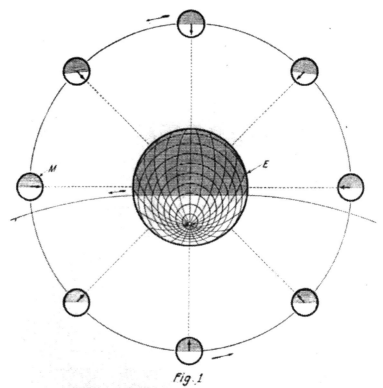

Fig. 1

It Is Well Known That the Moon, M., Always Turns the Same Face
Toward the Earth, E, as the Black Arrows Indicate. The Parallel
Rays From the Sun Illuminate the Moon In Its Successive Orbital
Positions as the Unshaded Semi-circles Indicate. Bearing This In
Mind, Do You Believe That the Moon Rotates on Its Own Axis?

It is well known since the discovery of Galileo that the moon, in
travelling thru space, always turns the same face towards the earth. This
is explained by stating that while passing once around its mother-planet
the lunar globe performs just one revolution on its axis. The spinning
motion of a heavenly body must necessarily undergo modifications in the
course of time, being either retarded by remittances internal or external,
or accelerated owing to shrinkage and other causes. An unalterable
rotational velocity thru all phases of planetary evolution is manifestly
impossible. What wonder, then, that at this very instant of its long
existence our satellite should revolve exactly so, and not faster or slower.
But many astronomers have accepted as a physical fact that such rotation
takes place. It does not, but only appears so; it is an illusion, a most
surprising one, too.

I will endeavor to make this clear by reference to Fig. 1, in which E represents the earth and M the moon. The movement thru space is such that the arrow, firmly attached to the latter, always occupies the position indicated with reference to the earth. If one imagines himself as looking

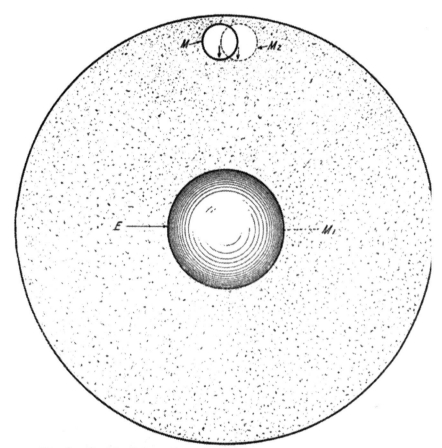

Fig. 2.—Tesla's Conception of the Rotation of the Moon, M, Around the Earth, E; the Moon, in This Demonstration Hypothesis, Being Considered as Embedded in a Solid Mass, M_1. If, As Commonly Believed, the Moon Rotates, This Would Be Equally True For a Portion of the Mass M_2, and the Part Common to Both Bodies Would Turn Simultaneously in "Opposite" Directions.

down on the orbital plane and follows the motion he will become convinced that the moon *does* turn on its axis as it travels around. But in this very act the observer will have deceived himself. To make the delusion complete let him take a washer similarly marked and supporting it rotatably in the center, carry it around a stationary object, constantly keeping the arrow pointing towards the latter. Tho to his bodily vision the disk will revolve on its axis, such movement does not exist. He can dispel the illusion at once by holding the washer fixedly while going around. He will now readily see that the supposed axial rotation is only apparent, the impression being produced by successive changes of position in space.

But more convincing proofs can be given that the moon does not, and cannot revolve on its axis. With this object in view attention is called to Fig. 2, in which both the satellite, M, and earth, E, are shown embedded in a solid mass, Mi, (indicated by stippling) and supposed to rotate so as to impact to the moon its normal translatory velocity. Evidently, if the lunar globe could rotate as commonly believed, this would be equally true of any other portion of mass M_1, as the sphere M_1, shown in dotted lines, and then the part common to both bodies would have to turn *simultaneously in opposite directions*. This can be experimentally illustrated in the manner suggested by using instead of one, two overlapping rotatable washers, as may be conveniently represented by circles M and M_2, and carrying them around a center as E, so that the plain and dotted arrows are always pointing towards the same center. No further argument is needed to demonstrate that the two gyrations cannot co-exist or even be pictured in the imagination and reconciled in a purely abstract sense.

The truth is, the so-called "axial rotation" of the moon is a phenomenon deceptive alike to the eye and mind and devoid of physical meaning. It has nothing in common with real mass revolution characterized by effects positive and unmistakable. Volumes have been written on the subject and many erroneous arguments advanced in support of the notion. Thus, it is reasoned, that if the planet did *not* turn on its axis it would expose the whole surface to terrestrial view; as only one-half is visible, it *must* revolve. The first statement is true but the logic of the second is defective, for it admits of only one alternative. The conclusion is not justified as the same appearance can also be produced in another way. The moon does rotate, not on its own, but about an axis passing thru the center of the earth, the true and only one.

The unfailing test of the spinning of a mass is, however, the existence of energy of motion. The moon is not possessed of such *vis viva*. If it were the case then a revolving body as M_1 would contain mechanical energy other than that of which we have experimental evidence. Irrespective of this so exact a coincidence between the axial and orbital periods is, in itself, immensely improbable for this is not the permanent condition towards which the system is tending. Any axial rotation of a mass left to itself retarded by forces external or internal, must cease. Even admitting its perfect control by tides the coincidence would still be miraculous. But when we remember that most of the satellites exhibit this peculiarity, the probability becomes infinitesimal.

Three theories have been advanced for the origin of the moon. According to the oldest suggested by the great German philosopher Kant, and developed by Laplace in his monumental treatise "Mécanique

Céleste," the planets have been thrown off from larger central masses by centrifugal force. Nearly forty years ago Prof. George H. Darwin in a masterful essay on tidal friction furnished mathematical proofs, deemed unrefutable, that the moon had separated from the earth. Recently this established theory has been attacked by Prof. T. J. J. See in a remarkable work on the "Evolution of the Stellar Systems," in which he propounds, the view that centrifugal force was altogether inadequate to bring about the separation and that all planets, including the moon, have come from the depths of space and have been captured. Still a third hypothesis of unknown origin exists which has been examined and commented upon by Prof. W. H. Pickering in "Popular Astronomy of 1907," and according to which the moon was torn from the earth when the later was partially solidified, this accounting for the continents which might not have been formed otherwise.

Undoubtedly planets and satellites have originated in both ways and, in my opinion, it is not difficult to ascertain the character of their birth. The following conclusions can be safely drawn:

1. A heavenly body thrown off from a larger one cannot rotate on its axis. The mass, rendered fluid by the combined action of heat and pressure, upon the reduction of the latter immediately stiffens, being at the same time deformed by gravitational pull. The shape becomes permanent upon cooling and solidification and the smaller mass continues to move about the larger one as tho it were rigidly connected to it except for pendular swings or librations due to varying orbital velocity. Such motion precludes the possibility of axial rotation in the strictly physical sense. The moon has never spun around as is well demonstrated by the fact that the most precise measurements have failed to show any measurable flattening in form.

2. If a planetary body in its orbital movement turns the same side towards the central mass this is a positive proof that it has been separated from the latter and is a true satellite.

3. A planet revolving on its axis in its passage around another cannot have been thrown off from the same but must have been captured.

II. The Fallacy of Franklin's Pointed Lightning-Rod

The display of atmospheric electricity has since ages been one of the most marvelous spectacles afforded to the sight of man. Its grandeur and power filled him with fear and For centuries he attributed lightning to agents god like and supernatural and its purpose in the scheme of this universe remained unknown to him. Now we have learned that the waters of the ocean are raised by the sun and maintained in the atmosphere delicately suspended, that they are waited to distant regions of the globe where electric forces assert themselves in upsetting the sensitive balance and causing precipitation, thus sustaining all organic life. There is every reason to hope that man will soon be able to control this life-giving flow of water and thereby solve many pressing problems of his existence.

Atmospheric electricity became of scientific interest in Franklin's time. Faraday had not yet announced his epochal discoveries in magnetic induction but static frictional machines were already generally used in physical laboratories. Franklin's powerful mind and once leaped to the conclusion that frictional and atmosphere electricity were identical. To our present view this inference appears obvious, but in his presence mere thought of it was little short of blasphemy. He investigated the phenomena and argued that if they were of the same nature then the clouds could be drained of their energy exactly as the ball of a static machine, and in 1749 he indicated in a published memoir how this could be done by the use of pointed metal rods.

The earliest trials were made by Dalibrand in France, but Franklin himself was the first to obtain a spark by using a kite, in June, 1752. When these atmospheric discharges manifest themselves today in our wireless station we feel annoyed and wish that they would stop, but to the man who discovered them they brought tears of joy.

The lightning conductor in its classical form was invented by Benjamin Franklin in 1755 and immediately upon its adoption proved a success to a degree. As usual, however, its virtues were often exaggerated. So, for instance, it was seriously claimed that in the city of Piatermaritzburg (capital of Natal, South Africa) no lightning strokes occurred after the pointed rods were installed, altho the storms were as frequent as before. Experience has shown that just the opposite is true. A modern city like

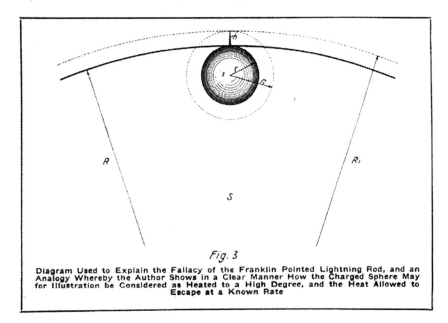

Fig. 3

Diagram Used to Explain the Fallacy of the Franklin Pointed Lightning Rod, and an Analogy Whereby the Author Shows in a Clear Manner How the Charged Sphere May for Illustration be Considered as Heated to a High Degree, and the Heat Allowed to Escape at a Known Rate

New York, presenting innumerable sharp points and projections in good contact with the earth, is struck much more often than equivalent area of land. Statistical records, carefully compiled and published from time to time, demonstrate that the danger from lightning to property and life has been reduced to a small percentage by Franklin's invention, but the damage by fire amounts, nevertheless, to several million dollars annually. It is astonishing that this device, which has been in universal use for more than one century and a half, should be found to involve a gross fallacy in design and construction which impairs its usefulness and may even render its employment hazardous under certain conditions.

For explanation of this curious fact I may first refer to Fig. 3, in which s is a metallic sphere of radius r, such as the capacity terminal of a static machine, provided with a sharply pointed pin of length h, as indicated. It is well known that the latter has the property of quickly dissipating the accumulated charge into the air. To examine this action in the light of present knowledge we may liken electric potential to temperature. Imagine that sphere s is heated to T degrees and that the pin or metal bar is a perfect conductor of heat so that its extreme end is at the same temperature T. Then if another sphere of larger radius, v_1, is drawn about the first and the temperature along this boundary is T_1, it is evident that there will be between the end of the bar and its surrounding a difference of temperature $T - T_1$, which will determine the outflow of heat. Obviously, if the adjacent medium was not affected by the hot sphere this temperature difference would be greater and more heat would be given off. Exactly so in the electric system. Let q be the quantity of the charge, then the sphere — and owing to its great conductivity also the pin — will be at $\dfrac{q}{r}$ the potential. The medium around the point of the pin will be at the potential $\dfrac{q}{r} - \dfrac{q}{r+h}$ and, consequently, the difference $\dfrac{q}{r} - \dfrac{q}{r+h} = \dfrac{qh}{r(r+h)}$. Suppose now that a sphere S of much larger radius $R = nr$ is employed containing a charge Q this difference of potential will be, analogously $\dfrac{Qh}{R(R+h)}$. According to elementary principles of electro-statics the potentials of the two spheres s and S will be equal if $Q = nq$ in which case $\dfrac{Qh}{R(R+h)} = \dfrac{nqh}{nr(nr+h)} = \dfrac{qh}{r(nr+h)}$. Thus the difference of potential between the point of the pin and the medium around the same will be smaller in the ratio $\dfrac{r+h}{nr+h}$ when the large sphere is used. In many scientific tests and experiments this important observation has been disregarded with the result of causing serious errors. Its significance is that the behavior of the pointed rod entirely depends on the linear dimensions of the electrified body. Its quality to give off the charge may be entirely lost if the latter is very large. For this reason, all points or projections on the surface of a conductor of such vast dimensions as the earth would be quite ineffective were it not for other influences. These will be elucidated with reference to Fig. 4, in which our artist of the Impressionist school has emphasized Franklin's notion that his rod was drawing electricity from the clouds. If the earth were not surrounded by an atmosphere which is generally oppositely charged it would behave, despite all its irregularities of surface, like a polished sphere. But owing to the electrified masses of air and cloud the distribution is greatly modified. Thus in Fig. 4. the positive charge of the cloud induces in the earth an equivalent of opposite charge, the density at the surface of the latter diminishing with the cube of the distance from the static center of the cloud. A brush discharge is then formed at the point of the rod and the action Franklin anticipated takes place. In addition, the surrounding air is ionized and rendered conducting and, eventually, a bolt may hit the building or some other object in the vicinity. The virtue of the pointed end to dissipate the charge, which was uppermost in Franklin's mind is, however, infinitesimal. *Careful measurements show that it would take many years before the electricity stored in a single cloud of moderate site would be drawn off or neutralized thru suck a lightning conductor.* The grounded rod has the quality of rendering harmless most of the strokes it receives, tho occasionally the charge is diverted with damaging results. But, what is very important to note, it invites danger and hazard on account of the fallacy involved in its

Fig. 4. Tesla Explains the Fallacy of the Franklin Pointed Lightning Rod, Here Illustrated, and Shows that Usually Such a Rod Could Not Draw Off the Electricity in a Single Cloud in Many Years. The Density of the Dots Indicates the Intensity of the Charges.

design. The sharp point which was thought advantageous and indispensable to its operation, is really a defect detracting considerably from the practical value of the device. I have produced a much improved form of lightning protector characterized by the employment of a terminal of considerable area and large radius of curvature which makes impossible undue density of the charge and ionization of the air. (*These protectors act as quasi-repellents and so far have never been struck tho exposed a long time.* Their safety is experimentally demonstrated to greatly exceed that invented by Franklin. By their use property worth millions of dollars which is now annually lost, can be saved.

III. The Singular Misconception of the Wireless.

To the popular mind this sensational advance conveys the impression of a single invention but in reality it is an art, the successful practice of which involves the employment of a great many discoveries and improvements. I viewed it as such when I undertook to solve wireless problems and it is due to this fact that my insight into its underlying principles was clear from their very inception.

In the course of development of my induction motors it became desirable to operate them at high speeds and for this purpose I constructed alternators of relatively high frequencies. The striking behavior of the currents soon captivated my attention and in 1889 I started a systematic investigation of their properties and the possibilities of practical application. The first gratifying result of my efforts in this direction was the transmission of electrical energy thru *one wire* without return, of which I gave demonstrations in my lectures and addresses before several scientific bodies here and abroad in 1891 and 1892. During that period, while working with my oscillation transformers and dynamos of frequencies up to 200,000 cycles per second, the idea gradually took hold of me that the earth might be used in place of the wire, thus dispensing with artificial conductors altogether. The immensity of the globe seemed an unsurmountable obstacle but after a prolonged study of the subject I became satisfied that the undertaking was rational, and in my lectures before the Franklin Institute and National Electric Light Association early in 1893 I gave the outline of the system I had conceived. In the latter part of that year, at the Chicago World's Fair, I had the good fortune of meeting Prof. Helmholtz to whom I explained my plan, illustrating it with experiments. On that occasion I asked the celebrated physicist for an expression of opinion on the feasibility of the scheme. He stated unhesitatingly that it was practicable, provided I could perfect apparatus capable of putting it into effect but this, he anticipated, would be extremely difficult to accomplish.

I resumed the work very much encouraged and from that date to 1896 advanced slowly but steadily, making a number of improvements the chief of which was my system of *concatenated tuned circuits* and method of regulation, now universally adopted. In the summer of 1897 Lord Kelvin happened to pass thru New York and honored me by a visit to my laboratory where I entertained him with demonstrations in support of my wireless theory. He was fairly carried away with what he saw but, nevertheless, condemned my protect in emphatic terms, qualifying it as something impossible, "an illusion and a snare." I had expected his approval and was pained and surprised. But the next day he returned and gave me a better opportunity for explanation of the advances I had made and of the true principles underlying the system I had evolved. Suddenly he remarked with evident astonishment: "Then you are not making use of Hertz waves?" "Certainly not," I replied, *"these are radiations."* No energy could be economically transmitted to a distance by any such agency. In my system the process is one of *true conduction* which, theoretically, can be effected at the greatest distance without appreciable loss." I can never forget the magic change that came over the illustrious philosopher the moment he freed himself from that erroneous impression. The skeptic who would not believe was suddenly transformed into the warmest of supporters. He parted from me not only thoroughly convinced of the scientific soundness of the idea but strongly expressed his confidence in its success. In my exposition to him I resorted to the following mechanical analogues of my own and the Hertz wave system.

Imagine the earth to be a bag of rubber filled with water, a small quantity of which is periodically forced in and out of the same by means of a reciprocating pump, as illustrated. If the strokes of the latter are effected in intervals of more than one hour and forty-eight minutes, sufficient for the transmission of the impulse thru the whole mass, the entire bag will expand and contract and corresponding movements will be imparted to pressure gauges or movable pistons with the same intensity, irrespective of distance. By working the pump faster, shorter waves will be produced which, on reaching the opposite end of the bag, may be reflected and give rise to stationary nodes and loops, but in any case, the fluid being incompressible, its inclosure perfectly elastic, and the frequency of oscillations not very high, the energy will be economically transmitted and very little power consumed so long as no work is done in the receivers. This is a crude but correct representation of my wireless system in which, however, I resort to various refinements. Thus, for instance, the pump is made part of a resonant system of great inertia, enormously magnifying the force of the impressed impulses. The receiving devices are similarly conditioned and in this manner the amount of energy collected in them vastly increased.

The Hertz wave system is in many respects the very opposite of this. To explain it by analogy, the piston of the pump is assumed to vibrate to and fro at a terrific rate and the orifice thru which the fluid passes in and out of the cylinder is reduced to a small hole. There is scarcely any movement of the fluid and almost the whole work performed results in the production of radiant heat, of which an infinitesimal part is recovered in a remote locality. However incredible, it is true that the minds of some of the ablest experts have been from the beginning, and still are, obsessed by this monstrous idea, and so it comes that the true wireless art, to which I laid the foundation in 1893, has been retarded in its development for twenty years. This is the reason why the "statics" have proved unconquerable, why the wireless shares are of little value and why the Government has been compelled to interfere.

We are living on a planet of well-nigh inconceivable dimensions, surrounded by a layer of insulating air above which is a rarefied and conducting atmosphere (Fig. 5). This is providential, for if all the air were conducting the transmission of electrical energy thru the natural media would be impossible. My early experiments have shown that currents of high frequency and great tension readily pass thru an atmosphere but moderately rarefied, so that the insulating stratum is reduced to a small thickness as will be evident by inspection of Fig. 6, in which a part of the

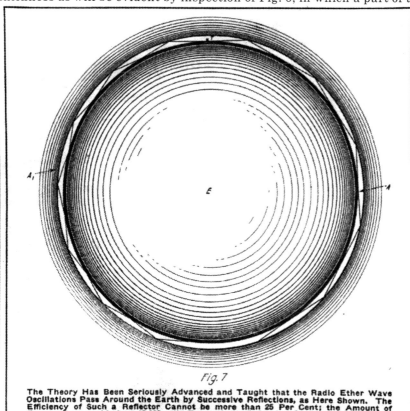

Fig. 7

The Theory Has Been Seriously Advanced and Taught that the Radio Ether Wave Oscillations Pass Around the Earth by Successive Reflections, as Here Shown. The Efficiency of Such a Reflector Cannot be more than 25 Per Cent; the Amount of Energy Recoverable in a 12,000-mile Transmission being but One Hundred and Fifteen Billionth Part of One Watt, with 1,000 Kilowatts at the Transmitter.

earth and its gaseous envelope is shown to scale. If the radius of the sphere is 12 ½," then the non-conducting layer is only 1/64" thick and it will be obvious that the Hertzian rays cannot traverse so thin a crack between two conducting surfaces for any considerable distance, without being absorbed. The theory has been seriously advanced that these radiations pass around the globe by *successive reflections*, but to show the absurdity of this suggestion reference is made to Fig. 7 in which this process is diagrammatically indicated. Assuming that there is no refraction, the rays, as shown on the right, would travel along the sides of a polygon drawn around the solid, and inscribed into the conducting gaseous boundary in which case the length of the side would be about 400 miles. As one-half the circumference of the earth is approximately 12,000 miles long there will be, roughly, thirty deviations. The efficiency of such a reflector cannot be more than 25 per cent, so that if none of the energy of the transmitter were lost in other ways, the part recovered would be measured by the fraction (¼)" Let the transmitter radiate Hertz waves at the rate of 1,000 kilowatts. Then about *one hundred and fifteen billionth part of one watt* is all that would be collected in a *perfect* receiver. In truth, the reflections would be much more numerous as shown on the left of the figure, and owing to this and other reasons, on which it is unnecessary to dwell, the amount recovered

MODE OF PROPAGATION OF THE CURRENT FROM THE TRANSMITTER THRU THE EARTH

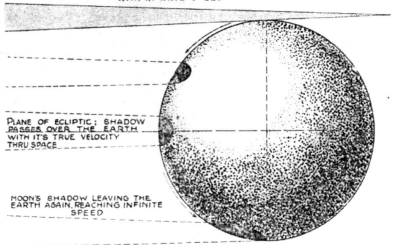

Fig. 8.—This Diagram Illustrates How, During a Solar Eclipse, the Moon's Shadow Passes Over the Earth With Changing Velocity, and Should Be Studied in Connection With Fig. 9. The Shadow Moves Downward With Infinite Velocity at First, Then With Its True Velocity Thru Space, and Finally With Infinite Velocity Again.

would be a vanishing quantity.

Consider now the process taking place in the transmission by the instrumentalities and methods of my invention. For this purpose attention is called to Fig. 8, which gives an idea of the mode of propagation of the current waves and is largely self-explanatory. The drawing represents a solar eclipse with the shadow of the moon just touching the surface of the earth at a point where the transmitter is located. As the shadow moves downward it will spread over the earth's surface, first with infinite and then gradually diminishing velocity until at a distance of about 6,000 miles it will attain its true speed in space. From there on it will proceed with increasing velocity, reaching infinite value at the opposite point of the globe. It hardly need be stated that this is merely an illustration and not an accurate representation in the astronomical sense.

The exact law will be readily understood by reference to Fig. 9, in which a transmitting circuit is shown connected to earth and to an antenna. The transmitter being in action, two effects are produced: Hertz waves pass thru the air, and a current traverses the earth. The former propagate with the speed of light and their energy is *unrecoverable* in the circuit. The latter proceeds with the speed varying as the cosecant of the angle which a radius drawn from any point under consideration forms with the axis of symmetry of the waves. At the origin the speed is infinite but gradually diminishes until a quadrant is traversed, when the velocity is that of light. From there on it again increases, becoming infinite at the antipole. Theoretically the energy of this current is *recoverable* in its entirety, in properly attuned receivers.

Some experts, whom I have credited with better knowledge, have for years contended that my proposals to transmit power without wires are sheer nonsense but I note that they are growing more cautious every day. The latest objection to my system is found in the cheapness of gasoline. These men labor under the impression that the energy flows in all directions and that,

Tesla's World-Wide Wireless Transmission of Electrical Signals, As Well As Light and Power, Is Here Illustrated In Theory, Analogy and Realization. Tesla's Experiments With 100 Foot Discharges At Potentials of Millions of Volts Have Demonstrated That the Hertz Waves Are Infinitesimal In Effect and Unrecoverable; the Recoverable Ground Waves of Tesla Fly "Thru the Earth". Radio Engineers Are Gradually Beginning to See the Light and That the Laws of Propagation Laid Down by Tesla Over a Quarter of a Century Ago Form the Real and True Basis of All Wireless Transmission To-Day.

therefore, only a minute amount can be recovered in any individual receiver. But this is far from being so. The power is conveyed in only one direction, from the transmitter to the receiver, and none of it is lost elsewhere. It is perfectly practicable to recover at any point of the globe energy enough for driving an airplane, or a pleasure boat or for lighting a dwelling. I am especially sanguine in regard to the lighting of isolated places and believe that a more economical and convenient method can hardly be devised. The future will show whether my foresight is as accurate now as it has proved heretofore.

Tesla Answers Mr. Manierre and Further Explains the Axial Rotation of the Moon

New York Tribune — Feb. 23, 1919

Sirs:

In your article of February 2, Mr. Charles E. Manierre, commenting upon my article in "The Electrical Experimenter" for February, which appeared in The Tribune of January 26, suggests that I give a definition of axial rotation.

I intended to be explicit on this point, as may be judged from the following quotation: "The unfailing test of the spinning of a mass is, however, the existence of energy of motion. The moon is not possessed of such vis viva." By this I meant that "axial rotation" is not simply "rotation upon an axis" as nonchalantly defined in dictionaries, but is circular motion in the true physical sense - that is, one in which half the product of the mass with the square of velocity is a definite and positive quantity.

The moon is a nearly spherical body, of a radius of about 1,081.5 miles, from which I calculate its volume to be approximately 5,300,216,300 cubic miles. Since its mean density is 3.27, one cubic foot of material composing it weighs close to 205 pounds. Accordingly, the total weight of the satellite is about 79,969,000,000, 000,000,000,000 and its mass 2,483,500,000,000,000,000 terrestrial short tons. Assuming that the moon does physically rotate upon its axis, it performs one revolution in 27 days 7 hours 43 minutes and 11 seconds, or 2,360,591 seconds. If, in conformity with mathematical principles, we imagine the entire mass concentrated at a distance from the center equal to two-fifths of the radius, then the calculated rotational velocity is 3.04 feet per second, at which the globe would contain 11,474,000,000,000,000,000 short foot tons of energy, sufficient to run 1,000,000, 000 horsepower for a period of 1,323 years. Now, I say that there is not enough energy in the moon to run a delicate watch.

In astronomical treatises usually the argument is advanced that "if the lunar globe did not turn upon its axis it would expose all parts to terrestrial view. As only a little over one-half is visible it must rotate." But this inference is erroneous, for it admits of one alternative. There are an infinite number of axes besides its own on each of which the moon might turn and still exhibit the same peculiarity.

I have stated in my article that the moon rotates about an axis, passing through the center of the earth, which is not strictly true, but does not vitiate the conclusions I have drawn. It is well known, of course, that the two bodies revolve around a common center of gravity which is at a distance of a little over 2,899 miles from the earth's center.

Another mistake in books on astronomy is made in considering this motion equivalent to that of a weight whirled on a string or in a sling. In the first place, there is an essential difference between these two devices though involving the same mechanical principle. If a metal ball attached to a string is whirled around and the latter breaks an axial rotation of the missile results which is definitely related in magnitude and direction to the motion preceding. By way of illustration: If the ball is whirled on the string clockwise, ten times a second, then when it flies off it will rotate on its axis twenty times a second, likewise in the direction of the clock. Quite different are the conditions when the ball is thrown from a sling. In this case a much more rapid rotation is imparted to it in the opposite sense. There is not true analogy to these in the motion of the moon. If the gravitational string, as it were, would snap, the satellite would go off in a tangent without the slightest swerving or rotation, for there is no momentum about the axis and, consequently, no tendency whatever to spinning motion.

Mr. Manierre is mistaken in his surmise as to what would happen if the earth were suddenly eliminated. Let us suppose that this would occur at the instant when the moon is in opposition. Then it would continue on its elliptical path around the sun, presenting to it steadily the face which was always exposed to the earth. If, on the other hand, the latter would disappear at the moment of conjunction, the moon would gradually swing around through 180 degrees and, after a number of oscillations, revolve again with the same face to the sun. In either case there would be no periodic changes, but eternal day and night, respectively, on the sides turned toward and away from the luminary.

Nikola Tesla

The Moon's Rotation

Electrical Experimenter — April, 1919

Since the appearance of my article entitled the "Famous Scientific Illusions" in your February issue, I have received a number of letters criticizing the views I expressed regarding the moon's "axial rotation." These have been partly answered by my statement to the *New York Tribune* of February 23, which allow me to quote:

In your issue of February 2, Mr. Charles E. Manierre, commenting upon my article in the *Electrical Experimenter* for February which appeared in the *Tribune* of January 26, suggests that I give a definition of axial rotation.

I intended to be explicit on this point as may be judged from the following quotation: "The unfailing test of the spinning of a mass is, however, the existence of *energy of motion*. The moon is not possessed of such *vis viva*." By this I meant that "axial rotation" is not simply "rotation upon an axis nonchalantly defined in dictionaries, but is a circular motion in the true physical sense—that is, one in which half the product of the mass with the square of velocity is a definite and positive quantity. The moon is a nearly spherical body, of a radius of about 1,087.5 miles, from which I calculate its volume to be approximately 5,300,216,300 cubic miles. Since its mean density is *327,* one cubic foot of material composing it weighs close on 205 lbs. Accordingly, the total weight of the satellite is about 79,969,000,000,000,000,000, and its mass 2,483,500,000,000,000,000 terrestrial short tons. Assuming that the moon does physically rotate upon its axis, it performs one revolution in 27 days, 7 hours, 43 minutes and 11 seconds, or 2,360,591 seconds. If, in conformity with mathematical principles, we imagine the entire mass concentrated at a distance from the center equal to two-fifths of the radius, then the calculated rotational velocity is 3.04 feet per second, at which the globe would contain 11,474,000,000,000,000,000 short foot tons of energy sufficient to run 1,000,000,000 horsepower for a period of 1,323 years. Now, I say, that there is not enough of that energy in the moon to run a delicate watch.

In astronomical treaties usually the argument is advanced that "if the lunar globe did not turn upon its axis it would expose all parts to terrestrial view. As only a little over one-half is visible it *must* rotate." But this inference is erroneous, for it only admits of one alternative. There are an infinite number of axis besides its own in each of which the moon might turn and still exhibit the same peculiarity.

I have stated in my article that the moon rotates about an axis passing thru the center of the earth, which is not strictly true, but it does not vitiate the conclusions I have drawn. It is well known, of course, that the two bodies revolve around a common center of gravity, which is at a distance of a little over 2,899 miles from the earth's center.

Another mistake in books on astronomy is made in considering this motion equivalent to that of a weight whirled on a string or in a sling. In the first place there is an essential difference between these two devices tho involving the same mechanical principle. If a metal ball, attached to a string, is whirled around and the latter breaks, an axial rotation of the missile results which is definitely related in magnitude and direction to the motion preceding. By way of illustration—if the ball is whirled on the string clockwise ten times per second, then when it flies off, it will rotate on its axis ten times per second, likewise in the direction of a clock. Quite different are the conditions when the ball is thrown from a sling. In this case a *much more rapid* rotation is imparted to it in the *opposite sense.* There is no true analogy to these in the motion of the moon. If *the gravitational string, as it were, would snap, the satellite would go off in a tangent without the slightest swerving or rotation, for there is no moment about the axis and, consequently, no tendency whatever to spinning motion.*

Mr. Manierre is mistaken in his surmise as to what would happen if the earth were suddenly eliminated. Let us suppose that this would occur at the instant when the moon is in *opposition.* Then it would continue on its elliptical path around the sun, presenting to it steadily the face which was always exposed to the earth. If, on the other hand, the latter would disappear at the moment of *conjunction,* the moon would gradually swing around thru 180° and, after a number of oscillations, revolve, again with the same face to the sun. In either case there would be no periodic changes but eternal day and night, respectively, on the sides turned towards, and away from, the luminary.

Some of the arguments advanced by the correspondents are ingenious and not a few comical. None, however, are valid.

One of the writers imagines the earth in the center of a circular orbital plate, having fixedly attached to its peripteral portion a disk-shaped moon, in frictional or geared engagement with another disk of the same diameter and freely rotatable on a pivot projecting from an arm entirely independent of the planetary system. The arm being held continuously parallel to itself, the pivoted disk, of course, is made to turn on its axis as the orbital plate is rotated. This is a well-known drive, and the rotation of the. pivoted disk is as palpable a fact as that of the orbital plate. But. the moon in this model only revolves about the center of the system *without the slightest angular displacement* on its own axis. The same is true of a cart-wheel to which this writer refers. So long as it advances on the earth's surface it turns on the axle in the true physical sense; when one of its spokes is always kept in a perpendicular position the wheel still *revolves* about the earth's center, *but axial rotation has ceased.* Those who think that it then still exists are laboring under an illusion.

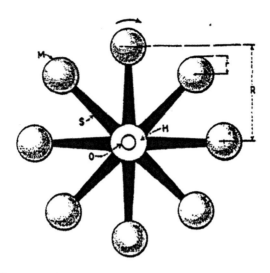

Fig. 1. — If You Still Think That the Moon Rotate on Its Axis, Look at This Diagram and Follow Closely the Successive Positions Taken by One of the Balls M While It is Rotated by a Spoke of the Wheel. Substitute Gravity for the Spoke and the Analogy Solves the Moon Rotation Riddle.

An obvious fallacy is involved in the following abstract reasoning. The orbital plate is assumed to gradually shrink, so that finally the centers of the earth and the satellite coincide when the latter revolves simultaneously about its own and the earth's axis. We may reduce the earth to a mathematical point and the distance between the two planets to the radius of the moon without affecting the system in principle, but a further diminution of the distance is manifestly absurd and of no bearing on the question under consideration.

In all the communications I have received, tho different in the manner of presentation, the successive changes of position in space are mistaken for axial rotation. So, for instance, a positive refutation of my arguments is found in the observation that the moon exposes all sides to other planets! It revolves, to be sure, but none of the evidences is a proof that it turns on its axis. Even the well-known experiment with the Foucault pendulum, altho exhibiting similar phenomena as on our globe, would merely demonstrate a motion of the satellite about *some* axis. The view I have advanced is *Not based on a theory* but on facts *demonstrable by experiment*. It is not a matter of *definition* as some would have it. *A Mass Revolving on its Axis Must Be Possessed of Momentum*. If it has none, there is no axial rotation, all appearances to the contrary notwithstanding.

A few simple reflections based on well established mechanical principles will make this clear. Consider first the case of two equal weights w and w_1, in Fig. 1, whirled about the center O on a string s as shown. Assuming the latter to break at a both weights will fly off on tangents to their circles of gyration, and, being animated with different velocities, they will rotate around their common center of gravity o. If the weights are whirled n times per second then the speed of the outer and the inner one will be, respectively, $V = 2 ® + r) n$ and $V_1 = 2 p (R—r) n$, and the difference $V—V_1 = 4$ p r n, will be the length of the circular path of the outer weight. Inasmuch, however, as there will be equalization of the speeds until the mean value is attained, we shall have $\dfrac{V - V_1}{2} = 2\pi rn = 2\pi rN$, N being the number of revolutions per second of the weights around their center of gravity. Evidently then, the weights continue to rotate at the original rate and in the same direction. I know this to be a fact from actual experiments. It also follows that a ball, as that shown in the figure, will behave in a similar manner for the two half-spherical masses can be concentrated at their centers of gravity and m and m_1, respectively, which will be at a distance from o equal to 3/8 r.

This being understood, imagine a number of balls M carried by as many spokes S radiating from a hub H, as illustrated in Fig. 2, and let this system be rotated n times per second around center O on frictionless bearings. A certain amount of work will be required to bring the structure to this speed, and it will be found that it equals exactly half the product of the masses with the square of the tangential velocity. Now if it be true that the moon rotates in reality on its axis *this must also hold good for each of the balls as it performs the same kind of movement.* Therefore, in imparting to the system a given velocity, energy must have been used up in the axial rotation of the balls. Let M be the mass of one of these and R the radius of gyration, then the rotational energy will be $E = \frac{1}{2}M (2pRn)^2$. Since for one complete turn of the wheel every ball makes one revolution on its axis, according to the prevailing theory, the energy of axial rotation of each ball will be $e = \frac{1}{2}M (2p r_1 n)^2$, r_1 being the radius of gyration about the axis and equal to 0.6325 r. We can use as large balls as we like, and so make e a considerable percentage of £ and yet, it is positively established by experiment that each of the rotating balls contain only the energy E, no power whatever being consumed in the supposed axial rotation, which is, consequently, wholly illusionary. Something even more interesting may, however, be stated. As I have shown before, a ball flying off will rotate at the rate of the wheel and in the same direction. But this

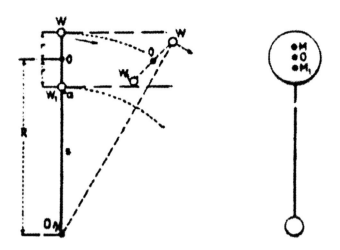

Fig. 2. — Diagram Illustrating the Rotation of Weights Thrown On By Centrifugal Force.

whirling motion, unlike that of a projectile, neither adds to, nor detracts from, the energy of the translatory movement which is exactly equal to the work consumed in giving to the mass the observed velocity.

From the foregoing it will be seen that in order to make one physical revolution on its axis the moon should have twice its present angular velocity, and then it would contain a quantity of stored energy as given in my above letter to the *New York Tribune,* on the assumption that the radius of gyration is 2/5 that of figure. This, of course, is uncertain, as the distribution of density in the interior is unknown. But from the character of motion of the satellite it may be concluded with certitude *that it is devoid of momentum about its axis.* If it be bisected by a plane tangential to the orbit, the masses of the two halves are inversely as the distances of their centers of gravity from the earth's center and, therefore, if the latter were to disappear suddenly, no axial rotation, as in the case of a weight thrown off, would ensue.

We believe the accompanying illustration and its explanation will dispel all doubts as to whether the moon rotates on its axis or not. Each of the balls, as M, depicts a different position of, and rotates exactly like, the moon keeping always the same face turned towards the center O, representing the earth.

But as you study this diagram, can you conceive that any of the balls turn on their axis? Plainly this is rendered physically impossible by the spokes. But if you are still unconvinced, Mr. Tesla's experimental proof will surely satisfy you. A body rotating on its axis must contain rotational energy. Now it is a fact, as Mr. Tesla shows, that no such energy is imparted to the ball as, for instance, to a projectile discharged from a gun.

It is therefore evident that the moon, in which the gravitational attraction is substituted for a spoke, cannot rotate on its axis or, in other words, contain rotational energy. If the earth's attraction would suddenly cease and cause it to fly off in a tangent, the moon would have no other energy except that of translatory movement, and it would not spin like the ball.—Editor.

Signals to Mars Based on Hope of Life on Planet
New York Herald — Sunday, Oct. 12, 1919

The idea that other planets are inhabited by intelligent beings might be traced to the very beginnings of civilization. This, in itself, would have little significance, for many of the ancient beliefs had their origin in ignorance, fear or other motives - good or evil, and were nothing more than products of untrained or tortured imagination. But when a conception lives through ages in the minds, growing stronger and stronger with increasing knowledge and intellectual development, it may be safely concluded that there is a solid truth underlying the instinctive perception. The individual is short lived and erring; man, relatively speaking, is imperishable and infallible. Even the positive evidences of the sense and the conclusions of science must be hesitatingly accepted when they are directed against the testimony of the entire body of humanity and the experience of centuries.

Modern investigation has disclosed the fact that there are other worlds, situated much the same as ours, and that organic life is bound to develop wherever there is heat, light and moisture. We know now that such conditions exist on innumerable heavenly bodies. In the solar system, two of these are particularly conspicuous — Venus and Mars. The former is, in many respects like the earth and must undoubtedly be the abode of some kind of life, but as to this we can only conjecture, for the surface is hidden from our view by a dense atmosphere. The latter planet can be readily observed and its periodic changes, which have been exhaustively studied by the late Percival Lowell, are a strong argument in support of the supposition that it is populated by a race vastly superior to ours in the mastery of the forces of nature

If such be the case then all that we can accomplish on this globe is of trifling importance as compared with the perfection of means putting us in possession of the secrets they must have discovered in their struggle against merciless elements. What a tragedy it would be were we to find some day that this wonderful people had finally met its inevitable fate and that all the precious intelligence they might have and, perhaps, had tried to convey to us, was lost. But although scientific research during the last few decades has given substance to the traditional belief, no serious attempt to establish communication could have been made until quite recently for want of proper instrumentalities.

Light Ray Project.

Long ago it was proposed to employ rays of light for this purpose and a number of men of science had devised specific plans which were discussed in the periodicals from time to time. But a careful examination shows that none of them is feasible, even on the assumption that the interplanetary space is devoid of gross matter, being filled only with a homogeneous and inconceivably tenuous medium called the ether. The tails of comets and other phenomena, however, would seem to disprove the theory, so that the successful exchange of signals by that kind of agency is very improbable.

While we can clearly discern the surface of Mars, it does not follow that the reverse is true. In perfect vacuum, of course, a parallel beam of light would be ideally suited for the transmission of energy in any amount for, theoretically, it could pass through infinite distance without any diminution of intensity. Unfortunately, this as well as other forms of radiant energy are rapidly absorbed in traversing the atmosphere.

It is possible that a magnetic force might be produced on the earth sufficient to bridge the gap of 50,000,000 miles and, in fact, it has been suggested to lay a cable around the globe with the object of magnetizing it. But certain electrical observations I made in studying terrestrial disturbances prove conclusively that there can not be much iron or other magnetic bodies in the earth beyond the insignificant quantity in the crust. Everything indicates that it is virtually a ball of glass and it would require many energizing turns to produce perceptible effects at great distance in this manner. Moreover, such an undertaking would be costly and, on account of the low speed of the current through the cable, the signaling would be extremely slow.

The Miracle Performed.

Such was the state of things until twenty years ago when a way was found to perform this miracle. It calls for nothing more than a determined effort and a feat in electrical engineering which, although difficult, is certainly realizable.

In 1899 I undertook to develop a powerful wireless transmitter and to ascertain the mode in which the waves were propagated through the earth. This was indispensable in order to apply my system intelligently for commercial purposes and, after careful study, I selected the high plateau of Colorado (6,000 feet above sea level) for the plant which I erected in the first part of that year. My success in overcoming the technical difficulties was greater than I had expected and in a few months I was able to produce electrical actions comparable to, and in a certain sense surpassing those of lightning. Activities of 18,000,000 horsepower were readily attained and I frequently computed the intensity of the effect in remote localities. During my experiments there, Mars was at a relatively small distance from us and, in that dry and rarefied air, Venus appeared so large and bright that it might have been mistaken for one of those military signaling lights. Its observation prompted me to calculate the energy transmitted by a powerful oscillator at 50,000,000 miles, and I came to the conclusion that it was sufficient to exert a noticeable influence on a delicate receiver of the kind I was, in the meanwhile, perfecting.

My first announcements to this effect were received with incredulity but merely because the potencies of the instrument I had devised were unknown. In the succeeding year, however, I designed a machine for a maximum activity of 1,000,000,000 horsepower which was partly constructed on Long Island in 1902 and would have been put in operation but for reverses and the fact that my project was too far in advance of the time.

It was reported at that period that my tower was intended for signaling to Mars, which was not the case, but it is true that I made a special provision for rendering it suitable to experiments in that direction. For the last few years there has been such a wide application of my wireless transmitter that experts have become, to an extent, familiar with its possibilities, and, if I am not mistaken, there are very few "doubting Thomases" now. But our ability to convey a signal across the gulf separating us from our neighboring planets would be of no avail if they are dead and barren or inhabited by races still undeveloped. Our hope that it might be different rests on what the telescope has revealed, but not on this alone.

Vast Power Found

In the course of my investigations of terrestrial electrical disturbances in Colorado I employed a receiver, the sensitiveness of which is virtually unlimited. It is generally believed that the so-called audion excels all others in this respect and Sir Oliver Lodge is credited with saying that it has been the means of achieving wireless telephony and transforming atomic energy. If the news is correct that scientist must have been victimized by some playful spirits with whom he is communicating. Of course, there is no conversion of atomic energy in such a bulb and many devices are known which can be used in the art with success.

My arrangements enable me to make a number of discoveries, some of which I have already announced in technical periodicals. The conditions under which I operated were very favorable for no other wireless plant of any considerable power existed and the effects t observed were thereafter due to natural causes, terrestrial or cosmic. I gradually learned how to distinguish in my receiver and eliminate certain actions and on one of these occasions my ear barely caught signals coming in regular succession which could not have been produced on the earth, caused by any solar or lunar action or by the influence of Venus, and the possibility that they might have come from Mars flashed upon my mind. In later years I have bitterly regretted that I yielded to the excitement of ideas and pressure of business instead of concentrating all my energies on that investigation.

The time is ripe now to make a systematic study of this transcending problem, the consummation of which may mean untold blessings to the human race. Capital should be liberally provided and a body of competent experts formed -to examine all the plans proposed and to assist in carrying out the best. The mere initiation of such a project in these uncertain and revolutionary times would result in a benefit which cannot be underestimated. In my early proposals I have advocated the application of fundamental mathematical principles for reaching the first elementary understanding. But since that time I have devised a plan akin to picture transmission through which knowledge of form could be conveyed and the barriers to the mutual exchange of ideas largely removed.

Success in Trials

Perfect success cannot be attained in any other way for we know only what we can visualize. Without perception of form there is not precise knowledge. A number of types of apparatus have been already invented with which transmission of pictures has been effected through the medium of wires, and they can be operated with equal facility by the wireless method. Some of these are of primitively simple construction. They are based on the employment of like parts which move in synchronism and transmit in this manner records, however complex. It would not require an extraordinary effort of the minds to hit upon this plan and devise instruments on this or similar principles and by gradual trials finally arrive at a full understanding.

The Herald of Sept. 24 contains a dispatch announcing that Prof. David Todd, of Amhurst College, contemplates an attempt to communicate with the inhabitants of Mars. The idea is to rise in a balloon to a height of about 50,000 feet with the manifest purpose of overcoming the impediments of the dense air stratum. I do not wish to comment adversely upon this undertaking beyond saying that no material advantage will be obtained by this method, for what is gained by height is offset a thousandfold by the inability of using powerful and complex transmitting and receiving apparatus. The physical stress and danger confronting the navigator at such an altitude are very great and he would be likely to lose his life or be permanently injured. In their recent record flights Roelfs and Schroeder have found that at a height of about six miles all their force was virtually exhausted. It would not have taken much more to terminate their careers fatally. If Prof. Todd wants to brave these perils he will have to provide special means of protection and these will be an obstacle to his observations. It is more likely, however, that he merely desires to look at the planet through a telescope in the hope of discerning something new. But it is by no means certain this instrument will be efficient under such conditions.

Interplanetary Communication
Electrical World — Sept. 24, 1921

To the Editors of the Electrical World:

There are countless worlds such as ours in the universe - planets revolving around their suns in elliptical orbits and spinning on their axes like gigantic tops. They are composed of the same elements and subject to the same forces as the earth. Inevitably at some period in their evolution light, heat and moisture are bound to be present, when inorganic matter will begin to run into organic forms. The first impulse is probably given by heliotropism; then other influences assert themselves, and in the course of ages, through continuous adjustment to the environment, automata of inconceivable complexity of structure result. In the workshop of nature these automatic engines are turned out in all essential respects alike and exposed to the same external influences.

The identity of construction and sameness of environment result in a concordance of action, giving birth to reason; thus intelligence, as the human, is gradually developed. The chief controlling agent in this process must be radiant energy acting upon a sense organ as the eye, which conveys a true conception of form. We may therefore conclude with certitude that, however constructively different may be the automata on other planets, their response to rays of light and their perceptions of the outside world must be similar to a degree so that the difficulties in the way of mutual understanding should not be insuperable.

Irrespective of astronomical and electrical evidences, such as have been obtained by the late Percival Lowell and myself, there is a solid foundation for a systematic attempt to establish communication with one of our heavenly neighbors, as Mars, which through some inventions of mine is reduced to a comparatively simple problem of electrical engineering. Others may scoff at this suggestion or treat it as a practical joke, but I have been in deep earnest about it ever since I made the first observations at my wireless plant in Colorado Springs from 1889 to 1900. Those who are interested in the subject may be referred to my articles in the *Century Magazine* of June, 1900, *Collier's Weekly* of Feb. 9, 1901, the *Harvard Illustrated Magazine* of March, 1907, the *New York Times* of May 23, 1909, and the *New York Herald* of Oct. 12, 1919.

At the time I carried on those investigations there existed no wireless plant on the globe other than mine, at least none that could produce a disturbance perceptible in a radius of more than a few miles. Furthermore, the conditions under which I operated were ideal, and I was well trained for the work. The arrangement of my receiving apparatus and the character of the disturbances recorded precluded the possibility of their being of terrestrial origin, and I also eliminated the influence of the sun, moon and Venus. As I then announced, the signals consisted in a regular repetition of numbers, and subsequent study convinced me that they must have emanated from Mars; this planet having been just then close to the earth.

Since 1900 I have spent a great deal of my time in trying to develop a thoroughly practical apparatus for the purpose and have evolved numerous designs. In one of these I find that an activity of 10,000,000,000 hp in effective wave energy could be attained. Assuming the most unfavorable conditions - namely, half-spherical propagation - then at a distance of 34,000,000 miles the energy rate would be about 1/730,000 hp per square mile, which is far more than necessary to affect a properly designed receiver. In fact, apparatus similar to that used in the transmission of pictures could be operated, and in this manner mathematical, geometrical and other accurate information could be conveyed.

I was naturally very much interested in reports given out about two years ago that similar observations had been made, but soon ascertained that these supposed planetary signals were nothing else than interfering undertones of wireless transmitters, and since I announced this fact other experts have apparently taken the same view. These disturbances I observed for the first time from 1906 to 1907. At that time they occurred rarely, but subsequently they increased in frequency. Every transmitter emits undertones, and these give by interference long beats, the wave length being anything from 50 miles to 300 or 400 miles. In all probability they would have been observed by many other experimenters if it were not so troublesome to prepare receiving circuits suitable for such long waves.

The idea that they would be used in interplanetary signaling by any intelligent beings is too absurd to be seriously commented upon. These waves have no suitable relation to any dimensions, physical constants or succession of events, such as would be naturally and logically considered in an intelligent attempt to communicate with us, and every student familiar with the fundamental theoretical principles will readily see that such waves would be entirely ineffective. The activity being inversely as the cube of the wave length, a short wave would be immensely more efficient as a means for planetary signaling, and we must assume that any beings who had mastered the art would also be possessed of this knowledge. On careful reflection I find, however, that the disturbances as reported, if they have been actually noted, cannot be anything else but forced vibrations of a transmitter and in all likelihood beats of undertones.

While I am not prepared to discuss the various aspects of this subject at length, I may say that a skillful experimenter who is in the position to expend considerable money and time will undoubtedly detect waves of about 25,470,000 m.

Nikola Tesla
New York City.

The following is a free preview of the Autobiography of
Nikola Telsa also available from Wilder Publications.
ISBN: 1-934451-77-0

My Inventions

My Early Life

The progressive development of man is vitally dependent on invention. It is the most important product of his creative brain. Its ultimate purpose is the complete mastery of mind over the material world, the harnessing of the forces of nature to human needs. This is the difficult task of the inventor who is often misunderstood and unrewarded. But he finds ample compensation in the pleasing exercises of his powers and in the knowledge of being one of that exceptionally privileged class without whom the race would have long ago perished in the bitter struggle against pitiless elements.

Speaking for myself, I have already had more than my full measure of this exquisite enjoyment, so much that for many years my life was little short of continuous rapture. I am credited with being one of the hardest workers and perhaps I am, if thought is the equivalent of labor, for I have devoted to it almost all of my waking hours. But if work is interpreted to be a definite performance in a specified time according to a rigid rule, then I may be the worst of idlers. Every effort under compulsion demands a sacrifice of life-energy. I never paid such a price. On the contrary, I have thrived on my thoughts.

In attempting to give a connected and faithful account of my activities in this series of articles which will be presented with the assistance of the Editors of the *Electrical Experimenter* and are chiefly addressed to our young men readers, I must dwell, however reluctantly, on the impressions of my youth and the circumstances and events which have been instrumental in determining my career.

Our first endeavors are purely instinctive, promptings of an imagination vivid and undisciplined. As we grow older reason asserts itself and we become more and more systematic and designing. But those early impulses, tho not immediately productive, are of the greatest moment and may shape our very destinies. Indeed, I feel now that had I understood and cultivated instead of suppressing them, I would have added substantial value to my bequest to the world. But not until I had attained manhood did I realize that I was an inventor.

This was due to a number of causes. In the first place I had a brother who was gifted to an extraordinary degree--one of those rare phenomena of mentality which biological investigation has failed to explain. His premature death left my parents disconsolate. We owned a horse which had been presented to us by a dear friend. It was a magnificent animal of Arabian breed, possessed of almost human intelligence, and was cared for and petted by the whole family, having on one occasion saved my father's life under remarkable circumstances. My father had been called one winter night to perform an urgent duty and while crossing the mountains, infested by wolves, the horse became frightened and ran away, throwing him violently to the ground. It arrived home bleeding and exhausted, but after the alarm was sounded immediately dashed off again, returning to the spot, and before the searching party were far on the way they were met by my father, who had recovered consciousness and remounted, not realizing that he had been lying in the snow for several hours. This horse was responsible for my brother's injuries from which he died. I witnessed the tragic scene and altho fifty-six years have elapsed since, my visual impression of it has lost none of its force. The recollection of his attainments made every effort of mine seem dull in comparison.

Anything I did that was creditable merely caused my parents to feel their loss more keenly. So I grew up with little confidence in myself. But I was far from being considered a stupid boy, if I am to judge from an incident of which I have still a strong remembrance. One day the Aldermen were passing thru a street where I was at play with other boys. The oldest of these venerable gentlemen—a wealthy citizen—paused to give a silver piece to each of us. Coming to me he suddenly stopped and commanded, "Look in my eyes." I met his gaze, my hand outstretched to receive the much valued coin, when, to my dismay, he said, "No, not much, you can get nothing from me, you are too smart." They used to tell a funny story about me. I had two old aunts with wrinkled faces, one of them having two teeth protruding like the tusks of an elephant which she buried in my cheek every time she kissed me. Nothing would scare me more than the prospect of being hugged by these as affectionate as unattractive relatives. It happened that while being carried in my mother's arms they asked me who was the prettier of the two. After examining their faces intently, I answered thoughtfully, pointing to one of them, "This here is not as ugly as the other."

Then again, I was intended from my very birth for the clerical profession and this thought constantly oppressed me. I longed to be an engineer but my father was inflexible. He was the son of an officer who served in the army of the Great Napoleon and, in common with his brother, professor of mathematics in a prominent institution, had received a military education but, singularly enough, later embraced the clergy in which vocation he achieved eminence. He was a very erudite man, a veritable natural philosopher, poet and writer and his sermons were said to be as eloquent as those of Abraham a Sancta-Clara. He had a prodigious memory and frequently recited at length from works in several languages. He often remarked playfully that if some of the classics were lost he could restore them. His style of writing was much admired. He penned sentences short and terse and was full of wit and satire. The humorous remarks he made were always peculiar and characteristic. Just to illustrate, I may mention one or two instances. Among the help there was a cross-eyed man called Mane, employed to do work around the farm. He was chopping wood one day. As he swung the axe my father, who stood nearby and felt very uncomfortable, cautioned him, "For God's sake, Mane, do not strike at what you are looking but at what you intend to hit." On another occasion he was taking out for a drive a friend who carelessly permitted his costly fur coat to rub on the carriage wheel. My father reminded him of it saying, "Pull in your coat, you are ruining my tire." He had the odd habit of talking to himself and would often carry on an animated conversation and indulge in heated argument, changing the tone of his voice. A casual listener might have sworn that several people were in the room.

Altho I must trace to my mother's influence whatever inventiveness I possess, the training he gave me must have been helpful. It comprised all sorts of exercises—as, guessing one another's thoughts, discovering the defects of some form or expression, repeating long sentences or performing mental calculations. These daily lessons were intended to strengthen memory and reason and especially to develop the critical sense, and were undoubtedly very beneficial.

My mother descended from one of the oldest families in the country and a line of inventors. Both her father and grandfather originated numerous implements for household, agricultural and other uses. She was a truly great woman, of rare skill, courage and fortitude, who had braved the storms of life and past thru many a trying experience. When she was sixteen a virulent pestilence swept the country. Her father was called away to administer the last sacraments to the dying and during his absence she went alone to the assistance of a neighboring family who were stricken by the dread disease. All of the members, five in number, succumbed in rapid succession. She bathed, clothed and laid out the bodies, decorating them with flowers according to the custom of the country and when her father returned he found everything ready for a Christian burial. My mother was an inventor of the first order and would, I believe, have achieved great things had she not been so remote from modern life and its multifold opportunities. She invented and constructed all kinds of tools and devices and wove the finest designs from thread which was spun by her. She even planted the seeds, raised the plants and separated the fibers herself. She worked indefatigably, from break of day till late at night, and most of the wearing apparel and furnishings of the home was the product of her hands. When she was past sixty, her fingers were still nimble enough to tie three knots in an eyelash.

There was another and still more important reason for my late awakening. In my boyhood I suffered from a peculiar affliction due to the appearance of images, often accompanied by strong flashes of light, which marred the sight of real objects and interfered with my thought and action. They were pictures of things and scenes which I had really seen, never of those I imagined. When a word was spoken to me the image of the object it designated would present itself vividly to my vision and sometimes I was quite unable to distinguish whether what I saw was tangible or not. This caused me great discomfort and anxiety. None of the students of psychology or physiology whom I have consulted could ever explain satisfactorily these phenomena. They seem to have been unique altho I was probably predisposed as I know that my brother experienced a similar trouble. The theory I have formulated is that the images were the result of a reflex action from the brain on the retina under great excitation. They certainly were not hallucinations such as are produced in diseased and anguished minds, for in other respects I was normal and composed. To give an idea of my distress, suppose that I had witnessed a funeral or some such nerve-racking spectacle. Then, inevitably, in the stillness of night, a vivid picture of the scene would thrust itself before my eyes and persist despite all my efforts to banish it. Sometimes it would even remain fixed in space tho I pushed my hand thru it. If my explanation is correct, it should be able to project on a screen the image of any object one conceives and make it visible. Such an advance would revolutionize all human relations. I am convinced that this wonder can and will be accomplished in time to come; I may add that I have devoted much thought to the solution of the problem.

To free myself of these tormenting appearances, I tried to concentrate my mind on something else I had seen, and in this way I would of ten obtain temporary relief; but in order to get it I had to conjure continuously new images. It was not long before I found that I had exhausted all of those at my command; my "reel" had run out, as it were, because I had seen little of the world--only objects in my home and the immediate surroundings. As I performed these mental operations for the second or third time, in order to chase the appearances from my vision, the remedy gradually lost all its force. Then I instinctively commenced to make excursions beyond the limits of the small world of which I had knowledge, and I saw new scenes. These were at first very blurred and indistinct, and would flit away when I tried to concentrate my attention upon them, but by and by I succeeded in fixing them; they gained in strength and distinctness and finally assumed the concreteness of real things. I soon discovered that my best comfort was attained if I simply went on in my vision farther and farther, getting new impressions all the time, and so I began to travel--of course, in my mind. Every night (and sometimes during the day), when alone, I would start on my journeys--see new places, cities and countries--live there, meet people and make friendships and acquaintances and, however unbelievable, it is a fact that they were just as dear to me as those in actual life and not a bit less intense in their manifestations.

This I did constantly until I was about seventeen when my thoughts turned seriously to invention. Then I observed to my delight that I could visualize with the greatest facility. I needed no models, drawings or experiments. I could picture them all as real in my mind. Thus I have been led unconsciously to evolve what I consider a new method of materializing inventive concepts and ideas, which is radically opposite to the purely experimental and is in my opinion ever so much more expeditious and efficient. The moment one constructs a device to carry into practice a crude idea he finds himself unavoidably engrossed with the details and defects of the apparatus. As he goes on improving and

reconstructing, his force of concentration diminishes and he loses sight of
the great underlying principle. Results may be obtained but always at the
sacrifice of quality.

My method is different. I do not rush into actual work. When I get an
idea I start at once building it up in my imagination. I change the
construction, make improvements and operate the device in my mind. It
is absolutely immaterial to me whether I run my turbine in thought or
test it in my shop. I even note if it is out of balance. There is no difference
whatever, the results are the same. In this way I am able to rapidly
develop and perfect a conception without touching anything. When I have
gone so far as to embody in the invention every possible improvement I
can think of and see no fault anywhere, I put into concrete form this final
product of my brain. Invariably my device works as I conceived that it
should, and the experiment comes out exactly as I planned it. In twenty
years there has not been a single exception. Why should it be otherwise?
Engineering, electrical and mechanical, is positive in results. There is
scarcely a subject that cannot be mathematically treated and the effects
calculated or the results determined beforehand from the available
theoretical and practical data. The carrying out into practice of a crude
idea as is being generally done is, I hold, nothing but a waste of energy,
money and time.

My early affliction had, however, another compensation. The incessant
mental exertion developed my powers of observation and enabled me to
discover a truth of great importance. I had noted that the appearance of
images was always preceded by actual vision of scenes under peculiar and
generally very exceptional conditions and I was impelled on each occasion
to locate the original impulse. After a while this effort grew to be almost
automatic and I gained great facility in connecting cause and effect. Soon
I became aware, to my surprise, that every thought I conceived was
suggested by an external impression. Not only this but all my actions
were prompted in a similar way. In the course of time it became perfectly
evident to me that I was merely an automaton endowed with power of
movement, responding to the stimuli of the sense organs and thinking and
acting accordingly. The practical result of this was the art of telautomatics
which has been so far carried out only in an imperfect manner. Its latent
possibilities will, however, be eventually shown. I have been since years
planning self-controlled automata and believe that mechanisms can be
produced which will act as if possessed of reason, to a limited degree, and
will create a revolution in many commercial and industrial departments.

I was about twelve years old when I first succeeded in banishing an
image from my vision by wilful effort, but I never had any control over the
flashes of light to which I have referred. They were, perhaps, my
strangest experience and inexplicable. They usually occurred when I
found myself in a dangerous or distressing situation, or when I was
greatly exhilarated. In some instances I have seen all the air around me
filled with tongues of living flame. Their intensity, instead of diminishing,

increased with time and seemingly attained a maximum when I was about twenty-five years old. While in Paris, in 1883, a prominent French manufacturer sent me an invitation to a shooting expedition which I accepted. I had been long confined to the factory and the fresh air had a wonderfully invigorating effect on me. On my return to the city that night I felt a positive sensation that my brain had caught fire. I saw a light as tho a small sun was located in it and I past the whole night applying cold compressions to my tortured head. Finally the flashes diminished in frequency and force but it took more than three weeks before they wholly subsided. When a second invitation was extended to me my answer was an emphatic NO!

These luminous phenomena still manifest themselves from time to time, as when a new idea opening up possibilities strikes me, but they are no longer exciting, being of relatively small intensity. When I close my eyes I invariably observe first, a background of very dark and uniform blue, not unlike the sky on a clear but starless night. In a few seconds this field becomes animated with innumerable scintillating flakes of green, arranged in several layers and advancing towards me. Then there appears, to the right, a beautiful pattern of two systems of parallel and closely spaced lines, at right angles to one another, in all sorts of colors with yellow-green and gold predominating. Immediately thereafter the lines grow brighter and the whole is thickly sprinkled with dots of twinkling light. This picture moves slowly across the field of vision and in about ten seconds vanishes to the left, leaving behind a ground of rather unpleasant and inert grey which quickly gives way to a billowy sea of clouds, seemingly trying to mold themselves in living shapes. It is curious that I cannot project a form into this grey until the second phase is reached. Every time, before falling asleep, images of persons or objects flit before my view. When I see them I know that I am about to lose consciousness. If they are absent and refuse to come it means a sleepless night.

To what an extent imagination played a part in my early life I may illustrate by another odd experience. Like most children I was fond of jumping and developed an intense desire to support myself in the air. Occasionally a strong wind richly charged with oxygen blew from the mountains rendering my body as light as cork and then I would leap and float in space for a long time. It was a delightful sensation and my disappointment was keen when later I undeceived myself.

During that period I contracted many strange likes, dislikes and habits, some of which I can trace to external impressions while others are unaccountable. I had a violent aversion against the earrings of women but other ornaments, as bracelets, pleased me more or less according to design. The sight of a pearl would almost give me a fit but I was fascinated with the glitter of crystals or objects with sharp edges and plane surfaces. I would not touch the hair of other people except, perhaps, at the point of a revolver. I would get a fever by looking at a peach and if a piece of camphor was anywhere in the house it caused me the keenest discomfort. Even now I am not insensible to some of these upsetting impulses. When I drop little squares of paper in a dish filled with liquid, I always sense a peculiar and awful taste in my mouth. I counted the steps in my walks and calculated the cubical contents of soup plates, coffee cups and pieces of food--otherwise my meal was unenjoyable. All repeated acts or operations I performed had to be divisible by three and if I mist I felt impelled to do it all over again, even if it took hours.

Up to the age of eight years, my character was weak and vacillating. I had neither courage or strength to form a firm resolve. My feelings came in waves and surges and vibrated unceasingly between extremes. My wishes were of consuming force and like the heads of the hydra, they multiplied. I was oppressed by thoughts of pain in life and death and religious fear. I was swayed by superstitious belief and lived in constant dread of the spirit of evil, of ghosts and ogres and other unholy monsters of the dark. Then, all at once, there came a tremendous change which altered the course of my whole existence. Of all things I liked books the best. My father had a large library and whenever I could manage I tried to satisfy my passion for reading. He did not permit it and would fly into a rage when he caught me in the act. He hid the candles when he found that I was reading in secret. He did not want me to spoil my eyes. But I obtained tallow, made the wicking and cast the sticks into tin forms, and every night I would bush the keyhole and the cracks and read, often till dawn, when all others slept and my mother started on her arduous daily task. On one occasion I came across a novel entitled "Abafi" (the Son of Aba), a Serbian translation of a well known Hungarian writer, Josika. This work somehow awakened my dormant powers of will and I began to practice self-control. At first my resolutions faded like snow in April, but in a little while I conquered my weakness and felt a pleasure I never knew before--that of doing as I willed. In the course of time this vigorous mental exercise became second nature. At the outset my wishes had to be subdued but gradually desire and will grew to be identical. After years of such discipline I gained so complete a mastery over myself that I toyed with passions which have meant destruction to some of the strongest men. At a certain age I contracted a mania for gambling which greatly worried my parents. To sit down to a game of cards was for me the quintessence of pleasure. My father led an exemplary life and could not excuse the senseless waste of time and money in which I indulged. I had a strong resolve but my philosophy was bad. I would say to him, "I can stop whenever I please but is it worth while to give up that which I would purchase with the joys of Paradise?" On frequent occasions he gave vent to his anger and contempt but my mother was different. She understood the character of men and knew that one's salvation could only be brought about thru his own efforts. One afternoon, I remember, when I had lost all my money and was craving for a game, she came to me with a roll of bills and said, "Go and enjoy yourself. The sooner you lose all we possess the

better it will be. I know that you will get over it." She was right. I
conquered my passion then and there and only regretted that it had not
been a hundred times as strong. I not only vanquished but tore it from my
heart so as not to leave even a trace of desire. Ever since that time I have
been as indifferent to any form of gambling as to picking teeth.

During another period I smoked excessively, threatening to ruin my
health. Then my will asserted itself and I not only stopped but destroyed
all inclination. Long ago I suffered from heart trouble until I discovered
that it was due to the innocent cup of coffee I consumed every morning.
I discontinued at once, tho I confess it was not an easy task. In this way
I checked and bridled other habits and passions and have not only
preserved my life but derived an immense amount of satisfaction from
what most men would consider privation and sacrifice.

After finishing the studies at the Polytechnic Institute and University
I had a complete nervous breakdown and while the malady lasted I
observed many phenomena strange and unbelievable.

My First Efforts At Invention

I shall dwell briefly on these extraordinary experiences, on account of their possible interest to students of psychology and physiology and also because this period of agony was of the greatest consequence on my mental development and subsequent labors. But it is indispensable to first relate the circumstances and conditions which preceded them and in which might be found their partial explanation.

From childhood I was compelled to concentrate attention upon myself. This caused me much suffering but, to my present view, it was a blessing in disguise for it has taught me to appreciate the inestimable value of introspection in the preservation of life, as well as a means of achievement. The pressure of occupation and the incessant stream of impressions pouring into our consciousness thru all the gateways of knowledge make modern existence hazardous in many ways. Most persons are so absorbed in the contemplation of the outside world that they are wholly oblivious to what is passing on within themselves.

The premature death of millions is primarily traceable to this cause. Even among those who exercise care it is a common mistake to avoid imaginary, and ignore the real dangers. And what is true of an individual also applies, more or less, to a people as a whole. Witness, in illustration, the prohibition movement. A drastic, if not unconstitutional, measure is now being put thru in this country to prevent the consumption of alcohol and yet it is a positive fact that coffee, tea, tobacco, chewing gum and other stimulants, which are freely indulged in even at the tender age, are vastly more injurious to the national body, judging from the number of those who succumb. So, for instance, during my student years I gathered from the published necrologues in Vienna, the home of coffee drinkers, that deaths from heart trouble sometimes reached sixty-seven per cent of the total. Similar observations might probably be made in cities where the consumption of tea is excessive. These delicious beverages superexcite and gradually exhaust the fine fibers of the brain. They also interfere seriously with arterial circulation and should be enjoyed all the more sparingly as their deleterious effects are slow and imperceptible. Tobacco, on the other hand, is conducive to easy and pleasant thinking and detracts from the intensity and concentration necessary to all original and vigorous effort of the intellect. Chewing gum is helpful for a short while but soon drains the glandular system and inflicts irreparable damage, not to speak of the revulsion it creates. Alcohol in small quantities is an excellent tonic, but is toxic in its action when absorbed in larger amounts, quite immaterial as to whether it is taken in as whiskey or produced in the stomach from sugar. But it should not be overlooked that all these are great eliminators assisting Nature, as they do, in upholding her stern but just law of the survival of the fittest. Eager reformers should also be mindful of the eternal perversity of mankind which makes the indifferent "laissez-faire" by far preferable to enforced restraint.

The truth about this is that we need stimulants to do our best work under present living conditions, and that we must exercise moderation and control our appetites and inclinations in every direction. That is what I have been doing for many years, in this way maintaining myself young in body and mind. Abstinence was not always to my liking but I find ample reward in the agreeable experiences I am now making. Just in the hope of converting some to my precepts and convictions I will recall one or two.

A short time ago I was returning to my hotel. It was a bitter cold night, the ground slippery, and no taxi to be had. Half a block behind me followed another man, evidently as anxious as myself to get under cover. Suddenly my legs went up in the air. In the same instant there was a flash in my brain, the nerves responded, the muscles contracted, I swung thru 180 degrees and landed on my hands. I resumed my walk as tho nothing had happened when the stranger caught up with me. "How old are you?" he asked, surveying me critically. "Oh, about fifty-nine," I replied. "What of it?" "Well," said he, "I have seen a cat do this but never a man." About a month since I wanted to order new eyeglasses and went to an oculist who put me thru the usual tests. He looked at me incredulously as I read off with ease the smallest print at considerable distance. But when I told him that I was past sixty he gasped in astonishment. Friends of mine often remark that my suits fit me like gloves but they do not know that all my clothing is made to measurements which were taken nearly 35 years ago and never changed. During this same period my weight has not varied one pound.

In this connection I may tell a funny story. One evening, in the winter of 1885, Mr. Edison, Edward H. Johnson, the President of the Edison Illuminating Company, Mr. Batchellor, Manager of the works, and myself entered a little place opposite 65 Fifth Avenue where the offices of the company were located. Someone suggested guessing weights and I was induced to step on a scale. Edison felt me all over and said: "Tesla weighs 152 lbs. to an ounce," and he guest it exactly. Stripped I weighed 142 lbs. and that is still my weight. I whispered to Mr. Johnson: "How is it possible that Edison could guess my weight so closely?" "Well," he said, lowering his voice. "I will tell you, confidentially, but you must not say anything. He was employed for a long time in a Chicago slaughter-house where he weighed thousands of hogs every day! That's why." My friend, the Hon. Chauncey M. Depew, tells of an Englishman on whom he sprung one of his original anecdotes and who listened with a puzzled expression but - a year later - laughed out loud. I will frankly confess it took me longer than that to appreciate Johnson's joke.

Now, my well being is simply the result of a careful and measured mode of living and perhaps the most astonishing thing is that three times in my youth I was rendered by illness a hopeless physical wreck and given up by physicians. More than this, thru ignorance and lightheartedness, I got into all sorts of difficulties, dangers and scrapes from which I extricated myself as by enchantment. I was almost drowned a dozen times; was nearly boiled alive and just mist being cremated. I was entombed, lost and frozen. I had hair-breadth escapes from mad dogs, hogs, and other wild animals. I past thru dreadful diseases and met with all kinds of odd mishaps and that I am hale and hearty today seems like a miracle. But as I recall these incidents to my mind I feel convinced that my preservation was not altogether accidental.

An inventor's endeavor is essentially lifesaving. Whether he harnesses forces, improves devices, or provides new comforts and conveniences, he is adding to the safety of our existence. He is also better qualified than the average individual to protect himself in peril, for he is observant and resourceful. If I had no other evidence that I was, in a measure, possessed of such qualities I would find it in these personal experiences. The reader will be able to judge for himself if I mention one or two instances. On one occasion, when about 14 years old, I wanted to scare some friends who were bathing with me. My plan was to dive under a long floating structure and slip out quietly at the other end. Swimming and diving came to me as naturally as to a duck and I was confident that I could perform the feat. Accordingly I plunged into the water and, when out of view, turned around and proceeded rapidly towards the opposite side. Thinking that I was safely beyond the structure, I rose to the surface but to my dismay struck a beam. Of course, I quickly dived and forged ahead with rapid strokes until my breath was beginning to give out. Rising for the second time, my head came again in contact with a beam. Now I was becoming desperate. However, summoning all my energy, I made a third frantic attempt but the result was the same. The torture of suppressed breathing was getting unendurable, my brain was reeling and I felt myself sinking. At that moment, when my situation seemed absolutely hopeless, I experienced one of those flashes of light and the structure above me appeared before my vision. I either discerned or guest that there was a little space between the surface of the water and the boards resting on the beams and, with consciousness nearly gone, I floated up, pressed my mouth close to the planks and managed to inhale a little air, unfortunately mingled with a spray of water which nearly choked me. Several times I repeated this procedure as in a dream until my heart, which was racing at a terrible rate, quieted down and I gained composure. After that I made a number of unsuccessful dives, having completely lost the sense of direction, but finally succeeded in getting out of the trap when my friends had already given me up and were fishing for my body.

That bathing season was spoiled for me thru recklessness but I soon forgot the lesson and only two years later I fell into a worse predicament. There was a large flour mill with a dam across the river near the city where I was studying at that time. As a rule the height of the water was only two or three inches above the dam and to swim out to it was a sport not very dangerous in which I often indulged. One day I went alone to the river to enjoy myself as usual. When I was a short distance from the masonry, however, I was horrified to observe that the water had risen and was carrying me along swiftly. I tried to get away but it was too late.

Luckily, tho, I saved myself from being swept over by taking hold of the wall with both hands. The pressure against my chest was great and I was barely able to keep my head above the surface. Not a soul was in sight and my voice was lost in the roar of the fall. Slowly and gradually I became exhausted and unable to withstand the strain longer. just as I was about to let go, to be dashed against the rocks below, I saw in a flash of light a familiar diagram illustrating the hydraulic principle that the pressure of a fluid in motion is proportionate to the area exposed, and automatically I turned on my left side. As if by magic the pressure was reduced and I found it comparatively easy in that position to resist the force of the stream. But the danger still confronted me. I knew that sooner or later I would be carried down, as it was not possible for any help to reach me in time, even if I attracted attention. I am ambidextrous now but then I was left-handed and had comparatively little strength in my right arm. For this reason I did not dare to turn on the other side to rest and nothing remained but to slowly push my body along the dam. I had to get away from the mill towards which my face was turned as the current there was much swifter and deeper. It was a long and painful ordeal and I came near to failing at its very end for I was confronted with a depression in the masonry. I managed to get over with the last ounce of my force and fell in a swoon when I reached the bank, where I was found. I had torn virtually all the skin from my left side and it took several weeks before the fever subsided and I was well. These are only two of many instances but they may be sufficient to show that had it not been for the inventor's instinct I would not have lived to tell this tale.

Interested people have often asked me how and when I began to invent. This I can only answer from my present recollection in the light of which the first attempt I recall was rather ambitious for it involved the invention of an apparatus and a method. In the former I was anticipated but the latter was original. It happened in this way. One of my playmates had come into the possession of a hook and fishing-tackle which created quite an excitement in the village, and the next morning all started out to catch frogs. I was left alone and deserted owing to a quarrel with this boy. I had never seen a real hook and pictured it as something wonderful, endowed with peculiar qualities, and was despairing not to be one of the party. Urged by necessity, I somehow got hold of a piece of soft iron wire, hammered the end to a sharp point between two stones, bent it into shape, and fastened it to a strong string. I then cut a rod, gathered some bait, and went down to the brook where there were frogs in abundance. But I could not catch any and was almost discouraged when it occurred to me to dangle the empty hook in front of a frog sitting on a stump. At first he collapsed but by and by his eyes bulged out and became bloodshot, he swelled to twice his normal size and made a vicious snap at the hook.

Immediately I pulled him up. I tried the same thing again and again and the method proved infallible. When my comrades, who in spite of their fine outfit had caught nothing, came to me they were green with envy. For a long time I kept my secret and enjoyed the monopoly but finally yielded to the spirit of Christmas. Every boy could then do the same and the following summer brought disaster to the frogs.

In my next attempt I seem to have acted under the first instinctive impulse which later dominated me - to harness the energies of nature to the service of man. I did this thru the medium of May-bugs - or June-bugs as they are called in America - which were a veritable pest in that country and sometimes broke the branches of trees by the sheer weight of their bodies. The bushes were black with them. I would attach as many as four of them to a crosspiece, rotably arranged on a thin spindle, and transmit the motion of the same to a large disc and so derive considerable "power." These creatures were remarkably efficient, for once they were started they had no sense to stop and continued whirling for hours and hours and the hotter it was the harder they worked. All went well until a strange boy came to the place. He was the son of a retired officer in the Austrian Army. That urchin ate May-bugs alive and enjoyed them as tho they were the finest blue-point oysters. That disgusting sight terminated my endeavors in this promising field and I have never since been able to touch a May-bug or any other insect for that matter.

After that, I believe, I undertook to take apart and assemble the clocks of my grandfather. In the former operation I was always successful but often failed in the latter. So it came that he brought my work to a sudden halt in a manner not too delicate and it took thirty years before I tackled another clockwork again. Shortly there after I went into the manufacture of a kind of pop-gun which comprised a hollow tube, a piston, and two plugs of hemp. When firing the gun, the piston was pressed against the stomach and the tube was pushed back quickly with both hands. The air between the plugs was compressed and raised to high temperature and one of them was expelled with a loud report. The art consisted in selecting a tube of the proper taper from the hollow stalks. I did very well with that gun but my activities interfered with the window panes in our house and met with painful discouragement. If I remember rightly, I then took to carving swords from pieces of furniture which I could conveniently obtain. At that time I was under the sway of the Serbian national poetry and full of admiration for the feats of the heroes. I used to spend hours in mowing down my enemies in the form of corn-stalks which ruined the crops and netted me several spankings from my mother. Moreover these were not of the formal kind but the genuine article.

I had all this and more behind me before I was six years old and had past thru one year of elementary school in the village of Smiljan where I was born. At this juncture we moved to the little city of Gospic nearby. This change of residence was like a calamity to me. It almost broke my heart to part from our pigeons, chickens and sheep, and our magnificent flock of geese which used to rise to the clouds in the morning and return from the feeding grounds at sundown in battle formation, so perfect that it would have put a squadron of the best aviators of the present day to shame. In our new house I was but a prisoner, watching the strange people I saw thru the window blinds. My bashfulness was such that I would rather have faced a roaring lion than one of the city dudes who strolled about. But my hardest trial came on Sunday when I had to dress up and attend the service. There I meet with an accident, the mere thought of which made my blood curdle like sour milk for years afterwards. It was my second adventure in a church. Not long before I was entombed for a night in an old chapel on an inaccessible mountain which was visited only once a year. It was an awful experience, but this one was worse. There was a wealthy lady in town, a good but pompous woman, who used to come to the church gorgeously painted up and attired with an enormous train and attendants. One Sunday I had just finished ringing the bell in the belfry and rushed downstairs when this grand dame was sweeping out and I jumped on her train. It tore off with a ripping noise which sounded like a salvo of musketry fired by raw recruits. My father was livid with rage. He gave me a gentle slap on the cheek, the only corporal punishment he ever administered to me but I almost feel it now. The embarrassment and confusion that followed are indescribable. I was practically ostracized until something else happened which redeemed me in the estimation of the community.

An enterprising young merchant had organized a fire department. A new fire engine was purchased, uniforms provided and the men drilled for service and parade. The engine was, in reality, a pump to be worked by sixteen men and was beautifully painted red and black. One afternoon the official trial was prepared for and the machine was transported to the river. The entire population turned out to witness the great spectacle. When all the speeches and ceremonies were concluded, the command was given to pump, but not a drop of water came from the nozzle. The professors and experts tried in vain to locate the trouble. The fizzle was complete when I arrived at the scene. My knowledge of the mechanism was nil and I knew next to nothing of air pressure, but instinctively I felt for the suction hose in the water and found that it had collapsed. When I waded in the river and opened it up the water rushed forth and not a few Sunday clothes were spoiled. Archimedes running naked thru the streets of Syracuse and shouting Eureka at the top of his voice did not make a greater impression than myself. I was carried on the shoulders and was the hero of the day.

Upon settling in the city I began a four-years' course in the so-called Normal School preparatory to my studies at the College or Real Gymnasium. During this period my boyish efforts and exploits, as well as troubles, continued. Among other things I attained the unique distinction of champion crow catcher in the country. My method of procedure was extremely simple. I would go in the forest, hide in the bushes, and imitate the call of the bird. Usually I would get several answers and in a short while a crow would flutter down into the shrubbery near me. After that all I needed to do was to throw a piece of cardboard to distract its attention, jump up and grab it before it could extricate itself from the undergrowth. In this way I would capture as many as I desired. But on one occasion something occurred which made me respect them. I had caught a fine pair of birds and was returning home with a friend. When we left the forest, thousands of crows had gathered making a frightful racket. In a few minutes they rose in pursuit and soon enveloped us. The fun lasted until all of a sudden I received a blow on the back of my head which knocked me down. Then they attacked me viciously. I was compelled to release the two birds and was glad to join my friend who had taken refuge in a cave.

In the schoolroom there were a few mechanical models which interested me and turned my attention to water turbines. I constructed many of these and found great pleasure in operating them. How extraordinary was my life an incident may illustrate. My uncle had no use for this kind of pastime and more than once rebuked me. I was fascinated by a description of Niagara Falls I had perused, and pictured in my imagination a big wheel run by the Falls. I told my uncle that I would go to America and carry out this scheme. Thirty years later I saw my ideas carried out at Niagara and marveled at the unfathomable mystery of the mind.

I made all kinds of other contrivances and contraptions but among these the arbalists I produced were the best. My arrows, when shot, disappeared from sight and at close range traversed a plank of pine one inch thick. Thru the continuous tightening of the bows I developed skin on my stomach very much like that of a crocodile and I am often wondering whether it is due to this exercise that I am able even now to digest cobble-stones! Nor can I pass in silence my performances with the sling which would have enabled me to give a stunning exhibit at the Hippodrome. And now I will tell of one of my feats with this antique implement of war which will strain to the utmost the credulity of the reader. I was practicing while walking with my uncle along the river. The sun was setting, the trout were playful and from time to time one would shoot up into the air, its glistening body sharply defined against a projecting rock beyond. Of course any boy might have hit a fish under these propitious conditions but I undertook a much more difficult task and I foretold to my uncle, to the minutest detail, what I intended doing. I was to hurl a stone to meet the fish, press its body against the rock, and cut it in two. It was no sooner said than done. My uncle looked at me almost scared out of his wits and exclaimed "*Vade retro Satanas!*" and it was a few days before he spoke to me again. Other records, how ever great, will be eclipsed but I feel that I could peacefully rest on my laurels for a thousand years.

CPSIA information can be obtained at www.ICGtesting.com
Printed in the USA
BVOW05s2208060314

346954BV00010B/216/P

In the following I shall consider three exceptionally interesting errors in the interpretation and application of physical phenomena which have for years dominated the minds of experts and men of science. 1. The Illusion of the Axial Rotation of the Moon. 2. The Fallacy of Franklin's Pointed Lightning-Rod. 3. The Singular Misconception of the Wireless.

Tesla is well known for his inventions, speeches and articles, many of which are readily available. Of those that were not widely available until now, this is perhaps his best known. Now you can read it for yourself, along with four essays on similar subjects.

ISBN 1-934451-99-1

90000

9 781934 451991

SPELLING WORDS for Year 4

2,000 WORDS EVERY CHILD SHOULD KNOW

KS2 English
Ages 8-9

STP

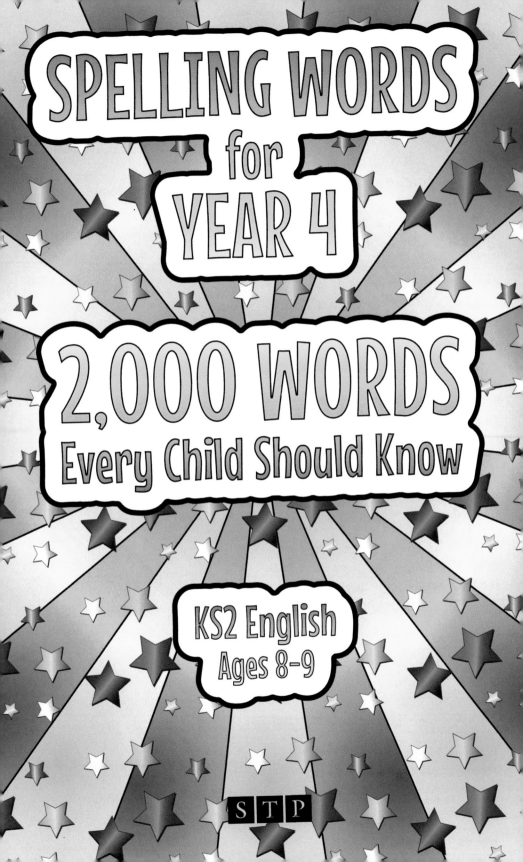

ABOUT THIS BOOK

Using a **fresh approach** to spellings lists, this illustrated collection of Spelling Words is designed **to make spelling fun** for children whilst ensuring they master essential spelling rules covered by the end of Year 4.

Containing **2,000** carefully selected **level-appropriate** words, this book is made up of **70** Themed Spellings Lists that

- Have **brightly-coloured illustrated backgrounds** and **engaging titles**
- Cover **loads of topics** that **actually interest children** such as Bonfire Night, the Internet, and vampires
- Relate to other **areas covered at school** including marine life, Roman gods & goddesses, and nutrition
- Target **key words that children overuse** (e.g. 'make', 'change', and 'real')
- Quietly introduce **specific areas of spelling** that children need to know (e.g. suffixes, homophones, doubling consonants, possessive plural nouns, and silent letters)
- Are made up of **25 to 30 words each**

HOW TO USE IT

All the **lists are self-contained**, so you can work through them **in order**, or, you can dip in to use them for **focused practice**. And, as these lists are themed, they are **also a useful resource** for a range of **writing projects and exercises**.

For your convenience, an **Index** to the **spelling rules, patterns, and themed areas** dealt with by each of the lists is included at the **back of the book** on page 40.

Published by STP Books
An imprint of Swot Tots Publishing Ltd
Kemp House
152-160 City Road
London EC1V 5NX

www.swottotspublishing.com

Text, design, illustrations and layout © Swot Tots Publishing Ltd
First published 2021

Swot Tots Publishing Ltd have asserted their moral right under the Copyright, Designs and Patents Act, 1988, to be identified as the author of this work.

Typeset, cover design, and inside concept design by Swot Tots Publishing Ltd.

British Library Cataloguing-in-Publication Data. A catalogue record for this book is available from the British Library.

ISBN 978-1-912956-40-1

CONTENTS

CONTENTS Cont.

Happy Endings I

accusation	conversation	inspiration
admiration	declaration	intonation
adoration	determination	modernization
aspiration	dispensation	organisation
civilization	dramatization	perspiration
colonization	examination	preparation
combination	exploration	restoration
condensation	imagination	sensation
configuration	improvisation	synchronization
conjuration	inclination	transfiguration

WWW

blog	hit	podcaster
blogging	home page	podcasting
bookmark	livestream	search engine
broadband	meme	spam
browser	message board	upload
chatroom	newsgroup	web address
cookie	offline	webcam
domain name	online	webpage
download	paywall	website
email	podcast	Wi-Fi

Silent, But Deadly I

high	align	gnome
bright	assign	gnu
fight	benign	phlegm
flight	malign	though
light	resign	through
sight	sign	
campaign	gnarled	
foreign	gnash	
reign	gnat	
sovereign	gnaw	

Subject To Change

alter	rectify	rework
amend	redo	transfigure
convert	refashion	transform
correct	refine	upgrade
emend	reform	vary
improve	rejig	
modify	remake	
morph	remodel	
mutate	revamp	
recast	revise	

Mis- The Misled

misadventure	misguided	misquote
misadvised	mishear	misread
misaligned	misinformed	misrule
misapplied	misinterpret	misshapen
misbehave	misjudge	misspell
misconduct	mislabel	mistake
misdeed	mislay	mistreat
misdiagnosis	misleading	mistrust
misdirect	mismatch	misunderstand
misfire	misplace	misuse

...Is A Person Who...

actor	governor	surveyor
administrator	inspector	survivor
commentator	instructor	traitor
conductor	inventor	translator
conqueror	investigator	tutor
constructor	legislator	
contractor	navigator	
decorator	sailor	
dictator	sculptor	
director	spectator	

Owww!

announced	doubt	mouse
bounce	encounter	pounce
boundary	foundation	roundabout
bounty	fountain	scoundrel
cloudy	hound	shroud
couch	loudest	slouch
countdown	lounge	soundless
countess	mound	spouse
countless	mount	thousand
county	mountains	warehouse

That Doesn't Look Right

accident	enough	weight
address	brought	weird
disappear	thought	complete
embarrass	decide	interest
grammar	recipe	mention
occur	beige	popular
successful	protein	purpose
suppose	receive	remember
tomorrow	their	strength
naughty	vein	tattoo

On The Hunt!

alligators	killer whales	snow leopards
buzzards	kites	tigers
condors	Komodo dragons	vultures
cougars	lions	whale sharks
crocodiles	lynxes	wolves
electric eels	moray eels	
falcons	ospreys	
grizzly bears	otters	
hawks	polar bears	
jaguars	pythons	

Make An Impact I

accelerate	elude	mortify
adore	empower	obliterate
appease	festoon	pester
bicker	flout	postpone
bulldoze	galvanize	rally
burden	grin	saunter
captivate	hone	tarnish
condemn	humiliate	thrive
deteriorate	intrude	thwart
dispel	lament	undertake

Business As Usu-al

animal	hospital	plural
capital	local	rival
coral	medal	rural
dual	metal	sandal
equal	moral	several
fatal	mural	terminal
formal	normal	total
general	oval	vertical
global	pedal	virtual
horizontal	petal	vital

What's Your Take?

bag	grip	secure
capture	hold	seize
catch	hook	snag
clasp	land	snap up
clench	latch onto	snare
cling to	lay hold of	snatch
clutch	nab	take
grab	nail	take hold of
grapple	net	trap
grasp	prise	wrest

Illegal To Irresponsible

illegal	inability	inhuman
illegible	inactive	inhumane
illiberal	incorrect	insecure
illiterate	indecisive	insoluble
illogical	indirect	invalid
immature	inedible	irrational
immortal	inequality	irregular
impatient	inexact	irrelevant
imperfect	inexplicable	irreparable
impossible	informal	irresponsible

Chic 'Ch'

brioche	chevalier	parachute
cartouche	chevron	pastiche
chalet	chic	pistachio
champagne	chiffon	quiche
chandelier	chute	ricochet
charade	cliché	
château	crochet	
chauffeur	gouache	
chef	machine	
chemise	moustache	

King Arthur & Co.

King Arthur	Camelot	joust
Merlin	Avalon	knights
Queen Guinevere	Excalibur	ladies
Sir Lancelot	Round Table	lance
Morgan Le Fay	Arthurian	lords
Lady of the Lake	betrayal	quests
Sir Gawain	chivalry	retinue
Sir Galahad	gallantry	romance
Mordred	giants	sword
Uther Pendragon	honour	tournaments

Joining Forces

afternoon	footprints	rattlesnake
backfire	headhunter	skyscraper
bathroom	inkblot	sunbeam
blackberry	keepsake	sunflower
cardboard	lighthouse	superpower
caveman	litterbug	timetable
deadline	mainstream	toothbrush
egghead	pinhole	touchdown
everywhere	postcard	underground
eyebrow	rainbow	upstairs

Ex Marks The Spot!

exceed	context	annex
excellent	dexterous	apex
excite	dyslexia	complex
exclamation	hexagon	convex
exercise	hypertext	flex
experience	lexicon	ibex
experiment	nexus	index
expert	plexus	perplex
extend	textile	vex
extremity	texture	vortex

Itchy Feet

ankle boots	galoshes	plimsolls
biker boots	golf shoes	running shoes
clogs	gym shoes	sandals
court shoes	half-boots	slingbacks
cowboy boots	high heels	slippers
Doc Martens	hobnail boots	sneakers
espadrilles	lace-ups	snowshoes
flats	loafers	stilettos
flip-flops	moccasins	tennis shoes
football boots	mules	Wellington boots

Grrr-ammar!

determiner	personal	past
noun	relative	present
verb	conjunction	future
modal verb	root word	simple
adverb	prefix	progressive
adjective	suffix	perfect
comparative	adverbial	active voice
superlative	phrase	passive voice
preposition	clause	direct speech
pronoun	subordinate	indirect speech

Let's Play...

backgammon	ludo	solitaire
battleships	Mastermind	speed chess
chequers	Monopoly	Trivial Pursuit
chess	nine men's morris	Twister
Chinese chequers	Operation	Yahtzee
Cluedo	Parcheesi	
Connect 4	Pictionary	
dominoes	Risk	
draughts	Scrabble	
Jenga	snakes and ladders	

Make, Not Break

assemble	fashion	prepare
author	forge	produce
build	form	raise
compose	formulate	rear
construct	generate	shape
craft	invent	
create	manufacture	
design	model	
devise	mould	
fabricate	originate	

Silent, But Deadly II

alms	chalk	yolky
balm	stalk	colonel
calm	talk	salmon
calming	talking	Norfolk
calf	walk	Suffolk
calves	walker	
half	could	
halves	should	
palm	folks	
palmed	yolk	

As The Crow Flies

chaffinch	magpie	sparrow
chough	moorhen	starling
emu	nightingale	thrush
flamingo	nightjar	woodpecker
hummingbird	partridge	wren
jackdaw	peacock	
jay	pheasant	
kite	raven	
kiwi	rook	
lark	seagull	

Happy Endings II

advantageous	studious	notorious
courageous	various	oblivious
outrageous	victorious	obvious
beauteous	curious	odious
bounteous	delirious	previous
envious	devious	religious
furious	hilarious	serious
glorious	imperious	tedious
luxurious	ingenious	hideous
mysterious	luminous	courteous

Remember, Remember...

Guy Fawkes	commemoration	Parliament
Guido Fawkes	confession	plotters
Robert Catesby	conspiracy	rebellion
James VI & I	effigy	scaffold
Gunpowder Plot	execution	treason
Palace of Westminster	explosion	
House of Lords	explosives	
November	fireworks	
assassination	gunpowder	
bonfire	memorial	

Double Trouble

forbidden	acquitted	catering
forbidding	acquitting	credited
concurred	emitting	crediting
concurring	submitted	creditor
conferred	submitting	opened
conferring	banqueted	opener
deferred	banqueter	opening
deferring	banqueting	poisoned
deterred	catered	poisoner
deterring	caterer	poisoning

That's Cheating!

cheat	fake	misinform
con	falsify	mislead
cozen	fiddle	misrepresent
crib	fleece	misstate
deceive	fudge	short-change
defraud	gull	swindle
distort	hoodwink	take in
doctor	hustle	tamper with
double-cross	lead on	twist
dupe	misguide	warp

All Over The Shop

top	minus	beneath
bottom	within	beyond
beginning	without	during
middle	beside	into
end	next to	past
right	around	
left	round	
up	amid	
down	amongst	
plus	against	

Perfect? Prefect?

perceive
percentage
performance
perfume
perhaps
peril
perimeter
periscope
perish
permanent

permit
persist
percussion
personify
persuade
precise
predatory
predict
preen
prefer

prelude
premium
president
pressure
prestige
presto
pretence
prettify
prevail
prevent

Farmers' Market

beetroot
broccoli
Brussels sprouts
cauliflower
celery
chicory
chillis
coriander
courgettes
fennel

kale
leeks
lettuce
mushrooms
parsley
parsnip
radish
rocket
shallot
spring onion

squash
swede
sweetcorn
turnip
yam

Make An Impact II

ablaze	earnest	placid
adventurous	enduring	prickling
arid	faultless	quaking
bedraggled	genial	shrouded
beloved	glazed	threadbare
boisterous	mindless	tolling
charismatic	monotonous	tumultuous
convenient	noiseless	unflinching
decrepit	outspoken	woeful
deluxe	penniless	wondrous

Kingly 'K'

ka	keyboard	kinetic
kangaroo	keystroke	kingfisher
karate	khaki	kingliness
karma	kickstand	kingmaker
keen	kidnapper	kink
keg	kilogram	kiosk
kennel	kilometre	kitchenware
kerchief	kindle	kohl
kettle	kindness	kookaburra
kettledrum	kindred	kraken

C Is For Camel &...

camel
canary
caribou
carp
caterpillar
catfish
centipede
chameleon
cheetah
chicken

chihuahua
chimpanzee
chinchilla
chipmunk
cicada
civet
clam
clownfish
cobra
cockatoo

cockroach
collie
condor
cormorant
coyote
crab
crane
crocodile
crow
cuttlefish

AAAATCHOOO!

batch
blotch
crutch
despatch
etch
glitch
notch
patch
sketch
snitch

stitch
twitch
catchy
ditched
hatchet
itchy
ketchup
kitchen
satchel
scratchy

thatched
watchful
watchman
wretched
butcher
catcher
dispatcher
pitcher
stretcher
watcher

Hide & Seek!

hide	smuggle	fish for
bury	squirrel	go after
camouflage	stash	hunt (for)
cloak	suppress	investigate
conceal	veil	probe
disguise	seek	pursue
entomb	chase	scrutinize
mask	comb	search (for)
obscure	delve	sniff out
secrete	ferret out	track down

Head-Scratchers

accept	fort	rain
except	fought	rein
affect	grate	war
effect	great	wore
bite	hoard	weather
byte	horde	whether
boar	law	wood
bore	lore	would
farther	main	who's
father	mane	whose

That's Quite Handy

Allen keys	ladder	rollers
axe	mallet	sander
bolts	measuring tape	sandpaper
chisel	nails	saw
clamps	nuts	screwdriver
duct tape	paintbrushes	spanner
hammer	pincers	spirit level
hand drill	pincers	toolbox
handsaw	pliers	torch
insulating tape	power drill	wrench
	putty	

Happy Endings III

angrily	luckily	terribly
busily	merrily	wiggly
cheerily	sleepily	basically
clumsily	steadily	chaotically
easily	bubbly	dramatically
greedily	fiddly	frantically
happily	gently	thematically
heavily	giggly	duly
hungrily	simply	truly
lazily	squiggly	wholly

Sweet Tooth

angel food cake	cupcake	macaroons
Bakewell tart	Danish pastry	millefeuille
baklava	devil's food cake	muffin
Battenburg	doughnut	pancake
Black Forest gateau	drop scone	panettone
brownie	dumpling	profiterole
coffee kiss	éclair	rum baba
cream cake	flapjack	scone
creampuff	frangipane	stollen
crumpet	gingerbread	Swiss roll

Plague? Plaque?

analogue	prologue	critique
colleague	rogue	grotesque
dialogue	tongue	masque
epilogue	vague	mosque
fatigue	vogue	mystique
intrigue	antique	opaque
league	baroque	physique
meringue	boutique	plaque
morgue	cheque	technique
plague	clique	unique

Happy Times!

anniversary	festival	masked ball
ball	festivity	masquerade
birthday	fête	pageant
carnival	fiesta	parade
cavalcade	function	party
celebration	gala	procession
centenary	graduation	prom
dance	holy day	reunion
fair	jamboree	soirée
fancy dress ball	jubilee	street party

The Deep Blue Sea

algae	mackerel	seahorse
barnacle	manatee	seal
barracuda	narwhal	squid
coral reef	octopus	starfish
dolphin	oyster	stingray
dugong	plankton	swordfish
hermit crab	porpoise	tuna
jellyfish	prawn	turtle
limpet	sardine	walrus
lobster	sea urchin	whale

STYLISH Y

ally	encyclopaedia	rhyme
apply	hydrogen	rye
bypass	hygiene	sly
cyberspace	hype	spry
cycle	hyphen	spy
cypress	ply	style
defy	pry	thyme
deny	pylon	type
dye	python	typhoon
dynamic	reply	tyrant

Glug, Glug, Glug

chug	imbibe	swallow
consume	knock back	swig
down	lap	swill
drain	partake of	swill down
drink	polish off	wash down
drink up	quaff	
finish off	sip	
gulp	slurp	
gulp down	suck	
guzzle	sup	

CoNFoUNded CoNJUNCtioNS

after	nor	where
although	once	wherever
and	or	while
because	since	whilst
before	so	yet
but	that	
for	unless	
how	until	
if	when	
neither	whenever	

Written In The Stars

Aries	Aquarius	character
Taurus	Pisces	personality
Gemini	astrology	potential
Cancer	horoscope	prediction
Leo	zodiac	strengths
Virgo	constellations	traits
Libra	astrologer	weaknesses
Scorpio	star sign	elements
Sagittarius	celestial	ascendant
Capricorn	chart	descendant

What's A Person Who...

comedian	egalitarian	veterinarian
custodian	equalitarian	equestrian
guardian	grammarian	pedestrian
theologian	humanitarian	dietician
civilian	librarian	beautician
thespian	nonsectarian	paediatrician
antiquarian	parliamentarian	patrician
authoritarian	sectarian	statistician
disciplinarian	totalitarian	tactician
documentarian	vegetarian	technician

Silent, But Deadly III

apostle	thistle	soften
bristle	trestle	often
bustle	whistle	mistletoe
castle	wrestle	mortgage
epistle	christen	ballet
gristle	fasten	beret
jostle	glisten	duvet
nestle	hasten	gourmet
pestle	listen	rapport
rustle	moisten	valet

WHERE IN THE WORLD...?

Alexandria	Khartoum	Paris
Bangkok	Kinshasa	Rio de Janeiro
Beijing	Lagos	Riyadh
Chicago	Lahore	São Paulo
Delhi	Lima	Seoul
Dhaka	Madrid	Shanghai
Istanbul	Manila	Singapore
Jakarta	Mexico City	Tehran
Johannesburg	Moscow	Tokyo
Karachi	New York City	Toronto

Colourific!

attractive	glossy	tinged
bleached	glowing	varnished
blotchy	matt	vibrant
bold	mellow	vivid
brilliant	mottled	watery
colourful	neutral	
electric	pale	
eye-catching	pastel	
gaudy	radiant	
glittering	striking	

Are You Sure...Or Ture?

closure
composure
displeasure
enclosure
exposure
acupuncture
agriculture
architecture
conjecture
curvature

denture
expenditure
fixture
horticulture
juncture
legislature
miniature
mixture
moisture
nurture

pasture
portraiture
posture
puncture
restructure
rupture
scripture
suture
tincture
venture

FULL OF BEANS!

adzuki beans
black beans
black-eyed beans
borlotti beans
broad beans
butter beans
cannellini beans
chickpeas
edamame beans
fava beans

flageolet beans
French beans
garbanzos
green beans
haricot beans
kidney beans
lima beans
mung beans
navy beans
pinto beans

red beans
runner beans
snap beans
soya beans
white beans

Sssss...

censorship	cilantro	furnace
census	cinder	palace
cent	cinema	space
centaur	cinnamon	notice
centigrade	cipher	office
centimetre	citadel	flounce
centre	citizen	silence
centrepiece	citrus	commerce
centurion	city	force
century	civil	source

Lots 'N Lots 'N Lots

keyboards'	princesses'	boxes'
pavements'	armies'	foxes'
posters'	ladies'	ibexes'
railings'	trophies'	taxes'
windows'	elves'	cacti's
bicycles'	hooves'	fungi's
cobblestones'	wolves'	appendices'
engines'	buffaloes'	indices'
compasses'	echoes'	analyses'
dresses'	mosquitoes'	geese's

By Jupiter!

Aesculapius	Fortuna	Nox
Apollo	Hercules	Pluto
Aurora	Janus	Proserpina
Bacchus	Juno	Saturn
Bellona	Jupiter	Silvanus
Ceres	Luna	Somnus
Cupid	Mars	Venus
Diana	Mercury	Vesta
Faunus	Minerva	Victoria
Flora	Neptune	Vulcan

Clippety-Clop, Clippety-Clop

canter	frolic	scuttle
crawl	gallop	skip
creep	gambol	skitter
dive	glide	slither
dogtrot	leap	soar
flap	lollop	stampede
flit	lope	strut
flutter	prance	swoop
fly	prowl	trot
frisk	scamper	waddle

Antihero Or Superstar?

antibiotic	automaton	submarine
antibody	autopilot	submerge
antidote	interchange	submerse
antihero	interject	subset
antiseptic	interlude	superbug
antitheft	international	superimpose
autobiography	interrupt	superman
autocracy	intersect	supermarket
autograph	subdivide	superscript
automatic	subheading	superstar

Knock On Wood

African mahogany	fir	red cedar
ash	hardwood	red fir
aspen	hazel	rosewood
balsa	hickory	sandalwood
banyan	larch	spruce
beech	mahogany	sycamore
birch	maple	teak
cedar	oak	walnut
ebony	pine	willow
elm	poplar	yew

Happy Endings IV

clinician
electrician
magician
mathematician
musician
optician
physician
politician
comprehension
expansion

extension
fusion
suspension
tension
admission
commission
discussion
expression
impression
possession

submission
action
creation
donation
edition
election
equation
invention
option
reaction

Time For Food

breakfast
lunch
dinner
supper
afternoon tea
banquet
barbecue
beanfeast
brunch
buffet

cream tea
elevenses
feast
high tea
luncheon
picnic
smorgasbord
snack
tapas
tiffin

continental breakfast
full English breakfast
packed lunch
potluck
indoor picnic
school dinner
dinner party
TV dinner
four-course meal
midnight feast

It's All Make-Believe...

chimeric	fantastic	invented
chimerical	fantastical	legendary
concocted	fictional	made-up
dreamed up	fictitious	make-believe
dreamlike	fictive	mythic
fabled	figmental	mythical
fabricated	ideal	mythological
fabulous	illusory	non-existent
fanciful	imaginary	pretend
fantasied	imagined	unreal

The Real Deal

accurate	incontestable	real
actual	incontrovertible	recognisable
authentic	indisputable	substantiated
bona fide	irrefutable	undeniable
certain	lawful	undoubted
certified	legal	unmistakable
established	legitimate	unquestionable
genuine	original	validated
guaranteed	proper	verifiable
identifiable	proven	verified

What A Ra-ck-et!

background	racket	hammock
beckon	sickle	hemlock
bracket	socket	hijack
cricket	speckled	maverick
cuckoo	tacky	padlock
gecko	aback	paperback
jackal	attack	ransack
jockey	flock	shipwreck
mockery	gimmick	sundeck
necklace	gridlock	thwack

HEADS FIRST

archduke	maharajah	rani
caliph	maharani	satrap
doge	mikado	Shah
duke	nabob	sheikh
emir	nawab	shogun
emperor	oba	stadholder
empress	pasha	sultan
khan	pope	tsar
khedive	queen	vicereine
king	rajah	viceroy

...Er

coroner	anger	conquer
hacker	answer	deliver
hiker	laughter	flicker
jeweller	power	prosper
juggler	semester	register
officer	bitter	surrender
partner	clever	swagger
trader	sinister	trigger
usher	slender	whimper
worker	tender	whisper

D Is For Disaster

annihilation	decimation	obliteration
apocalypse	demolition	ruin
Armageddon	destruction	tragedy
bloodbath	devastation	upheaval
calamity	doomsday	wreckage
carnage	havoc	
catastrophe	meltdown	
collapse	mischance	
crash	misfortune	
débâcle	mishap	

Write On!

biographer
blogger
columnist
creative writer
diarist
dramatist
editor
essayist
ghostwriter
gossip columnist

historian
humorist
journalist
librettist
lyricist
novelist
philosopher
playwright
poet
reporter

satirist
screenwriter
scribe
scriptwriter
tweeter

CLOUDING OVER

ashen
bleak
colourless
darkened
darkish
dim
dingy
dismal
drab
dreary

dull
dusky
eclipsed
faded
foggy
gloomy
grey
hazy
leaden
lightless

misty
muddy
murky
muted
overcast
shadowy
sombre
soupy
sunless
unilluminated

V..V.. VAMPIRES!

alluring	horror	silver
coffin	invitation	stake
consecrated	legend	stamina
crucifix	malevolent	sunlight
crypt	menacing	supernatural
Dracula	nocturnal	superstition
fangs	Nosferatu	terror
folklore	paranormal	threshold
garlic	predator	Transylvania
graveyard	shroud	undead

THAT'S A PROPER WORD?!

aplomb	jetsam	whack-a-mole
cowabunga	kibosh	windbag
diddly-squat	moonstruck	wiseacre
doodah	mudslinging	wuthering
egad	pell-mell	yokel
flotsam	quibble	
frippery	rapscallion	
gasbag	sassafras	
helter-skelter	slipshod	
humdrum	umlaut	

INDEX

In the following entries, the letter 'A' refers to the upper list on the page, while 'B' refers to the lower one.

Made in the USA
Columbia, SC
25 April 2022

59423356R00024